SpringerBriefs in Optimization

For further volumes:
http://www.springer.com/series/8918

My T. Thai

Group Testing Theory in Network Security

An Advanced Solution

My T. Thai
Department of Computer and Information
Science and Engineering
University of Florida
Gainesville, FL 32611
USA
e-mail: mythai@cise.ufl.edu

ISSN 2190-8354 e-ISSN 2191-575X
ISBN 978-1-4614-0127-8 e-ISBN 978-1-4614-0128-5
DOI 10.1007/978-1-4614-0128-5
Springer New York Dordrecht Heidelberg London

Library of Congress Control Number: 2011938392

Mathematics Subject Classification (2010):90C31, 68U07, 62P30

Printed on acid-free paper

Springer is part of Springer Science+Business Media (www.springer.com)

To my parents!

Preface

The idea of group testing is to discover defective items in a large population with the minimum number of tests, where each test, is applied to a subset of items, instead of testing them one by one. Although this powerful theory has been applied to many fields such as medical testing, codes, and multi-access channel communication, its newly emerging application into network security has been only recently discovered. Similar to all the earlier applications of group testing, this new application requires modifications or renovations of the classical group testing models and algorithms, so as to overcome the obstacles of applying the theoretical models to practical scenarios in network security. For example, under which conditions can group testing be constructed and tested on wireless sensor networks?

This monograph presents several new challenges and provides new group testing-based solutions for advanced network security problems. In particular, Chap. 2 presents a solution for Denial-of-Service attacks on the Internet in which a size constraint group testing is required. That is, there is a limit on the number of items in each subset and on the number of total subsets. Chapter 3 provides a solution for reactive jamming attacks in wireless sensor networks. Since the group testing is performed on wireless sensor networks, a careful design of an interference-free group testing, where each test result does not intefere with each other, is required. A more advanced solution for more sophisticated reactive jamming attacks is discussed in Chap. 4. Specifically, Chap. 4 provides a randomized fault-tolerant group testing construction to reduce the computational cost, compared to that using irreducible polynomials on Galois Field. Several challenges in designing a group testing-based solution for advanced reactive jamming attacks are also discussed. A discussion on open problems and suggestions for new solutions for various network security problems are included in the last chapter. We hope that this book will encourage research on the many intriguing open questions and applications of group testing that still remain.

Gainesville, FL, May 2011 My T. Thai

Contents

Chapter 1
Group Testing Theory

Abstract In this chapter, we briefly present an overview of group testing, its basic theory, and construction. We also present a general framework on using group testing for several network security defense schemes as well as discuss new challenges.

1.1 Introduction

The combinatorial Group Testing (GT) technique was first proposed by Dorfman during WWII for testing blood samples [1]. The idea of GT is to discover defective items in a large population with the minimum number of tests where each test is applied to a subset of items, called pools, instead of testing them one by one. When the outcome of a group test is negative, then all samples in the pool are good. Otherwise, there exists at least one defective sample, also called positive (but we do not know which one) in the pool and further testing on them is necessary. Since then, GT has been applied to many areas such as medical testing, codes, and multi-access channel communication.

In general, GT can be classified into two types: sequential (also called adaptive) and non-adaptive. A sequential GT conducts the tests one by one by using the results of previous tests to determine the pool for the next test, thereby completing the test within several rounds. A non-adaptive GT completes the test within one round by conducting all tests simultaneously, thus output results of the previous tests cannot be used to design the latter test. By taking advantages of previous testing results, sequential group testing requires fewer tests in general and was frequently used in the design of group testing since the main goal of group testing, historically, is to minimize the number of such tests in identifying all the positive samples.

Very recently, GT has been introduced to the molecular biology research field such as the DNA sequencing, and thus emerging new requirements [2–4]. Although minimizing the number of tests is still very important, the time required to finish the whole testing procedure must be considered since each single test may take from

M. T. Thai, *Group Testing Theory in Network Security*, SpringerBriefs in Optimization,
DOI: 10.1007/978-1-4614-0128-5_1, © My T. Thai 2012

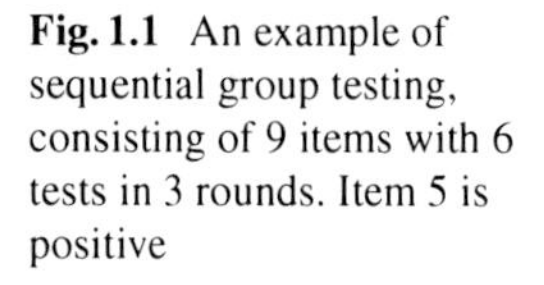

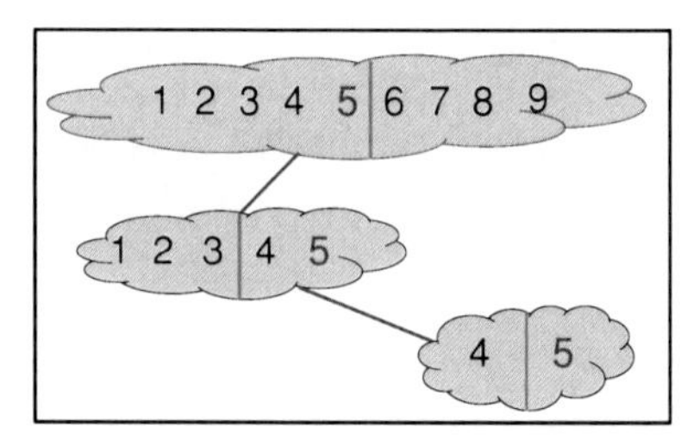

Fig. 1.1 An example of sequential group testing, consisting of 9 items with 6 tests in 3 rounds. Item 5 is positive

several hours to a day. The focus has then shifted to non-adaptive group testing where we can conduct all tests simultaneously, thus minimizing the testing time. However, constructing a non-adaptive group testing with the minimum number of pools is much more challenging. Determining the maximum number of tests required by an algorithm for given items and defected items is an open problem. Furthermore, it is not known whether this problem is NP-complete.

In this book, we focus more on non-adaptive GT than on sequential group testing as in the security setting, it is very important to detect the infected items as soon as possible. Therefore, we provide the basic theory and design of group testing, especially on the non-adaptive one in Sect. 1.2 The discussion of applying GT to security and its barriers to overcome are presented in Sect. 1.3.

1.2 Basic Theory and Design

1.2.1 Sequential Group Testing

For a sequential GT, at the end of each round, items in negative pools are identified as negative, while those in positive pools require to be further tested. Note that an item is identified as positive iff it is the only item in a positive pool. Thus sequential GT requires several testing rounds to finish a whole procedure. For instance, consider an example shown in Fig. 1.1, we have 9 items where item 5 is positive. Assume that we group these items into two pools, the test result of the first pool is positive whereas that of the second pool is negative. Thus further test on the first pool is required while none on the second. In total, we have to conduct 6 tests in three rounds.

Since each test divides the items into two disjoint sets, the worse-case number of tests required to test n items consisting of d positive items is at least $\left\lceil \log \binom{n}{d} \right\rceil$, which gives the well-known information theory bound.

There are several algorithms proposed to minimize the number of tests for sequential group testing. Normally, the number of positive items d is much smaller than n, thus if we can pool all negative items into one group, we can discard all those items within one test. But if the test outcome is positive, the smaller size of group is more preferable. Therefore, the determination of how large a group becomes a central goal

$$M = \begin{bmatrix} 1\,1\,1\,1\,0\,0\,0\,0\,0\,0 \\ 1\,0\,0\,0\,1\,1\,1\,0\,0\,0 \\ 0\,1\,0\,0\,1\,0\,0\,1\,1\,0 \\ 0\,0\,1\,0\,0\,1\,0\,1\,0\,1 \\ 0\,0\,0\,1\,0\,0\,1\,0\,1\,1 \end{bmatrix} \stackrel{\text{testing}}{\Longrightarrow} V = \begin{bmatrix} 0 \\ 1 \\ 0 \\ 0 \\ 1 \end{bmatrix}$$

Fig. 1.2 An example of non-adaptive group testing with 10 items where item 7th is positive. M is 1-disjunct. Vector V is a test outcome vector. Given M and V, we can identify item 7th as positive owing to the disjunctness of M. In this example, we conduct five tests in parallel, testing in one round

of most algorithms designed in the literature. Du and Hwang have dedicated a whole book discussing this topic [5].

1.2.2 Non-Adaptive Group Testing

The non-adaptive group testing consisting of t pools and n items including d positive ones can be represented by a $t \times n$ binary matrix M, where rows represent the pools and columns represent the items as shown in Fig. 1.2. An entry $M[i, j] = 1$ iff the ith pool contains the jth item; otherwise, $M[i, j] = 0$. Given this $M_{t \times n}$ matrix, a test result of these t pools can be represented by a t-dimensional column vector V, called the test outcome vector. V is a binary vector, in which 1-entry represents a positive outcome and 0- entry represents a negative one. A positive result indicates that at least one positive item exists within this pool, whereas a negative one means that all the items in the current pool are negative. That is, if $V[i] = 0$ then all items in row i of M are good; if $V[i] = 1$ then there exists at least one positive item in row i. Note that in the non-adaptive group testing, an item can be tested in several pools at the same time.

1.2.2.1 d-Disjunct Matrix

After conducting the test according to M and obtaining the test outcome V, we need to identify all d positive items from the population. This step is called decoding. In the non-adaptive group testing, a minimum requirement for M to be able to identify the set of positive items is that $U(D) \neq U(D')$ for $D \neq D'$, where $U(D)$ represents the Boolean sum of columns in D and D is a subset of d columns in M. For easy decoding, researchers in the literature have considered a d-disjunct property. M is called d-disjunct if no columns are contained in the Boolean sum (sometimes referred to as union) of any other d-columns. Note that 0 is said to be contained in 1 but not vice versa. Consider two columns $u = (0, 0, 1)^T$ and $v = (0, 1, 1)^T$. Then u is contained in v but v is not contained in u.

1.2.2.2 Decoding Algorithms

A decoding algorithm aims to identify the positive items by using the test outcome vector V and matrix M. Due to the d-disjunct properties, the decoding algorithm is quite simple. The following theorem suggests a decoding algorithm with an $O(tn)$ time complexity.

Lemma 1.1 *All unions of d distinct columns in a d-disjunct matrix are distinct*

Proof The proof is quite trivial, thus omitted. □

Theorem 1.1 *The number of items not in negative pools is always at most d*

Proof Note that an item does not appear in any negative pool iff its corresponding column is contained by the union of d positive columns. Therefore, the number of items not appearing in any negative pool is more than d iff there is at least a non-positive item whose column is contained by the d positive columns. But M is d-disjunct, hence Theorem 1.1 follows. □

Corollary 1.1 *The tests of negative outputs determine all negative items*

Therefore, we have the following decoding algorithm:

Algorithm 1 Decoding algorithm for d-disjunct matrix

1: **Input**: d-disjunct matrix M and test outcome vector V
2: **Output**: All positive items
3: Let S be the set of all n items
4: **for** each items $j \in S$ **do**
5:     **if** j is in a negative pool **then**
6:         $S = S \backslash j$
7:     **end if**
8: **end for**
9: Return S

1.2.2.3 Bounds on the Number of Tests

Let $t(d,n)$ denote the minimum number of rows required for a d-disjunct matrix with n columns; we now investigate the lower and upper bounds of this function.

The following result was attributed to Bassalygo by Dyachkov and Rykov [6].

Theorem 1.2 $t(d,n) \geq \min\{\binom{d+2}{2}, n\}$ *for a d-disjunct matrix with n columns.*

Proof We use the mathematical induction method to prove this theorem. The base case $n = 1$ is trivial. For the induction step, consider $d \geq 2$ and a d-disjunct matrix M with $t = t(d,n)$ rows and n columns. Let w denote a weight of a column in M, that is w of a column is equal to the number of 1s in that column. We consider the following two cases:

(i) Suppose that M has a column of $w \geq d + 1$. Since deleting a column and all rows intersecting it from a d-disjunct matrix yields a $(d - 1)$-disjunct matrix and base on the induction hypothesis, we have the following:

$$\begin{aligned} t(d,n) &\geq d + 1 + t(d-1, n-1) \\ &\geq d + 1 + \min\left\{\binom{d+1}{2}, n-1\right\} \\ &\geq \min\left\{\binom{d+1}{2}, n\right\} \end{aligned}$$

(ii) Suppose that M does not have such a column, then w of all columns are less than d. Let $c(w)$ denote the number of columns with weight w. A column is called isolated if there exists a row incident to it but not to any other column. It is easy to see that a non-isolated column in a d-disjunct matrix has weight at least $d+1$. Thus each column with $w \leq d$ is isolated. Hence,

$$n = \sum_{w=1}^{d} c(w) \leq t$$

The bound is thus proved.

Dyachkov et al. [7] have proved an upper bound of $t(d,n)$ using the random coding method as follows:

Theorem 1.3 *For a large n, $t(d,n) \leq (\log e) d^2 \log n (1 + o(1))$*

1.2.2.4 Matrix Construction

Numerous construction methods for d-disjunct matrix have been presented in [2, 8]. Among those solutions, we select to present Du's method, called *dDisjunct(n,d)*, as it has a better performance in terms of the size complexity (number of tests) [2]. The *dDisjunct(n, d)* algorithm takes n and d as the input, and outputs a d-disjunct matrix with n columns. The details of *dDisjunct (n,d)* algorithm is shown in Algorithm 2.

Basically, consider a finite field $GF(q)$ of order q. Suppose k satisfies $n \leq q^k$ and $t = d(k-1) + 1 \leq s \leq q$, Algorithm 2 first constructs a matrix A with s rows and n columns. Rows are indexed in the value of s, where columns are associated with n polynomials p_j for $j = 1, \ldots, n$ of degree $(k - 1)$ over $GF(q)$. The value of each cell in the matrix A is assigned as follows: For each row x and column p_j, we assign the value $p_j(x)$ over the finite field. To make sure we have enough polynomials to associate with n columns, we need $n \leq q^k$.

Based on matrix A, we construct a d-disjunct matrix M with $t = qs$ rows and n columns as follows. The columns are indexed in the polynomials of degree $k - 1$ while the rows are indexed in the ordered pairs in s values and in q values. For each row (x,y) in M, the entry $M[(x,y), p_j]$ is set to be 1 iff $p_j(x) = y$ in A. Otherwise, $M[(x,y), p_j] = 0$.

Algorithm 2 Construct a d-disjunct matrix (*dDisjunct(n,d)*)

```
 1: Consider a finite field GF(q), choose s,q,k satisfying:
 2:    t = d(k − 1) − 1 ≤ s ≤ q and n ≤ q^k
 3: ▷ Construct matrix A_{s×n}
 4: for x ∈ [0, s − 1] do
 5:    for all polynomials p_j of degree k − 1 do
 6:       A[x, p_j] ← p_j(x)
 7:    end for
 8: end for
 9: ▷Construct matrix M_{t×n}
10: for x ∈ [0, s − 1] do
11:    for y ∈ [0, q − 1] do
12:      for all polynomial p_j of degree k − 1 do
13:        if A[x, p_j] == y then
14:           M[(x, y), p_j] ← 1
15:         else
16:           M[(x, y), p_j] ← 0
17:         end if
18:       end for
19:      end for
20: end for
```

Theorem 1.4. (Correctness) *$M_{t\times n}$ obtained from Algorithm 2 is d-disjunct.*

Proof We first prove that matrix A has the following property: for any $(d+1)$ columns $p_0, p_1, \ldots, p_d$, there exists a row x such that $A[x, p_0]$ is not contained in the union of $A[x, p_j]$ for $j = 1, \ldots, d$. Assume that such a row x does not exist. Then for any row x_i, $p_0(x_i) = p_j(x_i)$ for some $j \in \{1, \ldots, d\}$. Since there are $d(k-1)+1$ rows, there exists a $j \in \{1, \ldots, d\}$ such that $p_0(x_i) = p_j(x_i)$ for at least k distinct rows. It follows that $p_0 = p_j$, a contradiction. Due to such a property of A and the construction of M based on A, it is easy to verify that M is d-disjunct. □

Theorem 1.5. (Performance) *The number of tests t(d,n) obtained from Algorithm 2 is*:

$$t(d,n) = \min\{(1+o(1))\frac{d^2\log^2 n}{\log^2(d\log n)}, n\}$$

Proof We need to choose q to have a positive integers k and t satisfying

$$\log_q n \le k \le \frac{s-1}{d} \le \frac{q-1}{d}$$

It suffices to choose q such that

$$\log_q n \le \frac{q-1}{d}$$

That implies,

$$n^d \le q^{q-1}$$

Let q_0 be the smallest number q satisfying the above condition. We claim that

$$q_0 \le (1+o(1))\frac{d\log_2 n}{\log_2(d\log_2 n)}$$

To prove this claim, in fact, we set

$$q^* = 1 + (1+h(d,n))\frac{2d\log_2 n}{\log_2(d\log_2 n)}$$

where

$$h(d,n) = \frac{\log_2\log_2(d\log_2 n)}{\log_2(d\log_2 n) - \log_2\log_2(d\log_2 n)}$$

Note that $h(d,n) = o(1) \ge 0$. Therefore, we have:

$$\begin{aligned}(q^*-1)\log_2 q^* &> (q^*-1)\log_2(q^*-1)\\ &\ge \frac{(1+h(d,n))d\log_2 n}{\log_2(d\log_2 n)}\log_2\frac{(1+h(d,n))d\log_2 n}{\log_2(d\log_2 n)}\\ &> d\log_2 n\end{aligned}$$

Therefore, it follows that $q_0 \le q^*$, and hence we have:

$$q_0 \le (1+o(1))\frac{d\log_2 n}{\log_2(d\log_2 n)}$$

This completes the proof. □

1.2.3 Error Tolerance

It is well known that there may exist some errors in experiments. The test may return some false negative or false positive results. In the false negative, the pool contains some positive items but the test result is negative due to some testing errors. Likewise, in the false positive, the pool contains all negative items. How do we correct these errors?

Let us consider a z-error correcting model. In this model, we assume that there is at most z errors in testing. After constructing a d-disjunct matrix M and obtaining an outcome vector V which consists of at most z errors, the z-error correcting model is still able to correct these errors in order to identify all the positive items.

z-**error correcting**. A matrix is said to be z-error correcting if the Hamming distance of any two unions of d columns is at least $2z+1$, where the Hamming distance of two column vectors is the number of different components between them.

Thus, in the group testing with error tolerance, we need to construct a d-disjunct matrix with z-error correcting, which is called (d,z)-disjunct matrix, defined as follows:

Definition 1.1 *(d,z)-disjunct matrix*: A $t \times n$ binary matrix M is (d, z)-disjunct if for any one column j and any other d columns $j_1, j_2, \ldots, j_d$, there exist $z+1$ rows $i_0, i_1, \ldots, i_z$ such that $M_{i_u j} = 1$ and $M_{i_u j_v} = 0$ for $u = 0, 1, \ldots, z$ and $v = 1, 2, \ldots, d$.

Theorem 1.6 *For every (d,z)-disjunct matrix, the Hamming distance between any two unions of d columns is at least* $2z+2$.

Proof This is straightforward by the definition of (d,z)-disjunct matrix. Let $C_1 \cup \cdots \cup C_d$ and $C_{1'} \cup \cdots \cup C_{d'}$ be two different unions of d columns. Assume $C_1 \notin \{C_{1'}, \ldots, C_{d'}\}$ and $C_{1'} \notin \{C_1, \ldots, C_d\}$. Then C_1 has at least $z+1$ 1-components not contained by $C_{1'} \cup \cdots \cup C_{d'}$, and likewise, $C_{1'}$ has at least $z+1$ 1-components not contained by $C_1 \cup \cdots \cup C_d$. Therefore, the Hamming distance between the two unions is at least $2z+2$. □

Decoding Algorithm The following theorem reveals a nice decoding algorithm for group testing with error tolerance.

Theorem 1.7 *Suppose a group testing is based on a (d,z)-disjunct matrix. If the number of errors is at most z, then the number of negative pools containing a positive item is always smaller than the number of negative pools containing a negative item.*

Proof Let i be a positive item and j be a negative item. Suppose there are m negative pools containing i. Since i is positive, the results of m pools must have some false negative. Hence, there are at most $z - m$ error tests turning a negative output to a positive output. Moreover, if no error exists, the number of negative pools containing j is at least $z+1$ due to (d,z)-disjunct. Hence the number of negative pools containing j is at least $(z+1) - (z-m) = m+1 > m$. □

From Theorem 1.7, we see that to decode d positive items from testing based on (d,z)-disjunct matrix, for each item i, we only need to count the number of negative pools containing i and select d smallest one. This runs in time $O(tn)$ and the decoding algorithm is presented in Algorithm 3.

Algorithm 3 Decoding algorithm for (d,z)-disjunct matrix

1: **Input**: (d,z)-disjunct matrix M and test result vector V
2: **Output**: All d positive items
3: $T = \emptyset$
4: **for** each item $i = 1 \ldots n$ **do**
5: $t(i) =$ number of negative pools containing i
6: $T = T \cup t(i)$
7: **end for**
8: Sort t_i in T in non-decreasing order
9: Return the first d smallest ones

We defer to present the construction of (d, z)-disjunct matrix in Chap. 4 in which we discuss more advanced solutions for network security.

1.3 Applications in Network Security

Despite many applications of group testing since WW II, it has only recently been applied to network security due to the work of Thai et al. [9]. Let us consider a simple attack scenario where we have n clients connecting to k servers and among n clients, we have d attackers. Once an attacker starts attacking a server, the resources of this server will be exhausted dramatically, thus it is not hard to identify which server is a victim. Therefore, a remaining question is to identify these d attackers from n clients as soon as possible. The key problem is how to group clients and assign them to different server machines in a sophisticated way, so that if any server is found under attack, we can immediately identify and filter the attackers out of its client set. Apparently, this problem resembles the group testing theory. A detection model based on GT can be constructed as follows. Consider a d-disjunct matrix $M_{t\times n}$, the clients can be mapped into the columns and servers into rows in M, where $M[i,j]=1$ if the requests from client j are distributed to server i. With regard to the test outcome column V, we have $V[i]=1$ iff server i is under attack from at least one attacker, but we cannot identify the attacker at once unless this server is handling only one client. Otherwise, if $V[i]=0$, all the clients assigned to server i are legitimate. The d attackers can then be captured by decoding the test outcome vector V and matrix M.

However, the detection model based on group testing is not trivial. Similar to all the earlier applications of GT, this new application to network security requires modifications of the classical GT model and algorithms, so as to overcome the obstacles of applying the theoretical models to practical scenarios. For example, the classical GT theory assumes that each pool can have as many items as needed and the number of pools for testing is unrestricted. However, with respect to our previous example, a server (equivalent to a pool in GT) cannot have infinite capacity, that is, it cannot serve infinite number of clients. In addition, there is a limit on a number of available servers for testing, thus we cannot have as many pools as possible.

Therefore, there must be some constraints on t and the total number of 1s on each row of M.

Network security is not limited to only network servers but also to other wireless networks, such as wireless sensor networks, where the test must be performed on the sensor nodes which have a very limited capacity. Thus from the designing of matrix M and performing the test to collecting the test outcomes become very challenging. In order to apply the GT into network security for a detection model, we need to handle the following important issues:

1. **Where to test:** As mentioned earlier, if testing is performed on the server (or virtual server), there must be a constraint on both t and the total number of 1s on each row of M. Thus a d-disjunct matrix M must satisfy these two constraints, thereby requiring new matrix construction algorithms which are described in Chap. 2. If testing is performed on a group of sensor nodes, we must consider the location of these nodes and their interferences. Thus we cannot freely group them together according to M. Chapter 3 provides a solution for this problem.
2. **How to test:** Depending on different applications, we have different ways of handling the testing. One of the major concerns is the indication of negative and positive results. For example, Chap. 2 discusses about a detection solution for Denial-of-Service (DoS) attack where testing is performed on virtual servers. In this case, we have to derive a dynamic attack threshold to indicate the server under attack based on the current server's load. If testing is done by sending some test message in wireless networks, then the schedule of transmitting this message and channels must be considered.
3. **How to collect results and decode:** Collecting the result is not as simple as testing in biology. In networks, we must design a method to send a result to either a central location to decode or must perform the decoding locally. If decoding is performed locally, we cannot simply use Algorithm 2 as we do not have all the test results of a whole network.
4. **Other system configuration:** From the system perspective, we have to consider the rest of the system configuration to finish a detection model. Assume that we have taken care of all the above three issues depending on different network models, attacker models, and victim models, we have to confront with the distribution of matrix M for mapping in a secured manner, the length of testing period, and other network characteristics.

The rest of this book discusses how to solve these four important issues one by one.

References

1. Dorfman R (1943) The detection of defective members of large populations. Ann Math Statist 14:436–440
2. Du D-Z, Hwang FK (2006) Pooling designs: group testing in molecular biology. World Scientific, Singapore

3. Thai MT, Znati T (2009) On the complexity and Approximation of non-unique probe selection using d-Disjunct Matrix. Journal of Combinatorial Optimization, special issues on Data Mining in Biomedicine 17(1):45–53
4. Thai MT, MacCallum D, Deng P, Wu W (2007) Decoding algorithms in pooling designs with inhibitors and fault tolerance. IJBRA 3(2):145–152
5. Du D-Z, Hwang FK (2000) Combinatorial group testing and its applications. World Scientific, Singapore
6. Dyachkov AG, Rykov VV (1983) A survey of superimposed code theory. Prob Control Inform Thy 12:229–242
7. Dyachkov AD, Rykov VV, Rachad AM (1989) Superimposed distance codes. Prob Control Inform Thy 18:237–250
8. Eppstein D, Goodrich MT, Hirschberg D (2005) Improved combinatorial group testing algorithms for real-world problem sizes, WADS. LNCS 3608, Springer, Berlin, pp 86–98
9. Thai MT, Xuan Y, Shin I, Znati T (2008) On detection of malicious users using group testing techniques. In: Proceedings of IEEE international conference on distributed computing systems (ICDCS), pp 206–213

Chapter 2
Size Constraint Group Testing and DoS Attacks

Abstract In this chapter, we introduce the first application of group testing in detecting application Denial-of-Service (DoS) attack , which aims at disrupting application service rather than depleting the network resource. This attack has emerged as one of the greatest threat to network services. Owing to its high similarity to legitimate traffic and much lower launching overhead than classic DoS attack, this new assault type cannot be efficiently detected or prevented by existing detection solutions. To identify application DoS attack, we present a novel group testing (GT)-based approach deployed on back-end servers, which not only offers a theoretical method to obtain short detection delay and low false positive/negative rate, but also provides an underlying framework against general network attacks. This new application requires a new class of group testing, called size constraint group testing.

2.1 Overview

One of the most critical problems in Internet security is the Denial-of-Service (DoS) attack , which aims to make a service unavailable to legitimate clients [1]. DoS attacks mainly abuse the network bandwidth around the Internet subsystems and degrade the quality of service by generating congestions at the network [1, 2]. However, with the boost in network bandwidth and application service types recently, the target of DoS attacks have shifted from network to server resources and application procedures themselves, forming a new application DoS attack [3].

By exploiting the flaws in application design and implementation, application DoS attacks exhibit three advantages over traditional DoS attacks which help evade normal detections: malicious traffic is always indistinguishable from normal traffic, adopting automated script to avoid the need for a large amount of "zombie" machines or bandwidth to launch the attack, much harder to be traced due to multiple redirections at proxies. According to these characteristics, the malicious traffic can be classified into legitimate-like requests of two cases: (1) at a high interarrival rate, (2) consuming

M. T. Thai, *Group Testing Theory in Network Security*, SpringerBriefs in Optimization,
DOI: 10.1007/978-1-4614-0128-5_2,

more service resources. We call these two cases "high-rate" and "high-workload" attacks, respectively.

Since malicious requests can be made arbitrarily similar to legitimate ones, accurately and efficiently distinguishing them is a challenging task for any defense mechanism. In addition, these attacks usually do not cause congestion at the network level, thus bypassing the network-based defense systems [1]. As a result, efficiently identifying the attackers in application DoS attacks, which is the focus of this chapter, becomes much more difficult.

Consequently, several network-based defense methods have tried to detect these attacks by controlling traffic volume or differentiating traffic patterns at the intermediate routers [4–9]. However, these approaches aim to defend at the network layer, i.e, differences in traffic patterns, traffic volume, which the application DoS attacks can bypass. Moreover, each traffic is inspected against the normal behaviors model, thereby increasing the time complexity. Because of that, detection and mitigation at the end-system of the victim servers have been proposed [3, 10, 11]. Among them the DDoS shield [3] and CAPTCHA-based defense [10] are the representatives of the two major techniques of system-based approaches: session validation based on legitimate behavior profile and authentication using human-solvable puzzles. In [3], each session is inspected against a normal behaviors model (i.e. interarrival time distribution) to detect a suspected session. Based on this suspicion level, a set of abnormal requests is enumerated. Note that the method keeps state for each session, which grows linearly with the number of clients. In [10], the service requires clients to solve puzzles and eliminates those who send a wrong solution at a high rate. However, this method is restricted to services with human clients who can solve puzzles. Plus, it introduces additional service delays for legitimate clients. It also checks each answer from clients one by one.

As the victim's resources will be exhausted, detecting application DoS attacks is always possible based on the performance of the attacked resources. Thus the key problem to identifying attackers lies in how to group clients together on each server so that if a server is under attack, we can quickly identify these attackers without examining each request. This problem resembles the group testing theory, therefore, the GT technique is uniquely suitable for the detection of application DoS attackers with high accuracy and efficiency. By utilizing the GT technique, attackers are detected based merely on the performance of attacked resources, without the need to keep track of attack signatures, tightly specify legitimate behavior, or examine each request one by one. Thus the GT-based defense scheme may overcome the limitations of current detection approaches.

The realization of this approach, however, raises several challenges. The classical GT theory assumes that each pool can have as many items as needed and the number of pools for testing is unrestricted. However, in order to provide real application services, servers cannot have infinite quantity or capacity, that is, the number of available pools is given and we cannot freely assign many clients/requests into a sever (pool). Thus constraints on these two parameters are required to complete the testing model, which we call Size Constraint Group Testing.

In a system viewpoint, the defense scheme is to embed multiple virtual servers within each physical back-end server, and map these virtual servers to the testing pools in GT, then assign clients into these pools by distributing their service requests to different virtual servers. By periodically monitoring some indicators (e.g., average responding time) for resource usage in each server, and comparing them with some dynamic thresholds, all the virtual servers can be judged as "safe" or "under attack". By means of the decoding algorithm of GT, all the attackers can be identified. Therefore, the biggest challenges of this method are threefold: (1) How to construct a testing matrix to enable prompt and accurate detection. (2) How to regulate the service requests to match the matrix, in a practical system. (3) How to establish proper thresholds for server source usage indicator to generate accurate test outcomes.

In the rest of this chapter, we first present the attacker and detection models to provide more information above the network security setting. The above three questions are addressed in Sects. 2.4 and 2.5.

2.2 Network System Models

2.2.1 DoS Attacker Model

The maximum destruction caused by the attacks includes the depletion of the application service resource at the server side, the unavailability of service access to legitimate user, and possible fatal system errors which require rebooting the server for recovery. Any malicious behaviors can be discovered by monitoring the service resource usage, based on dynamic value thresholds over the monitored objects. We also assume that application interface presented by the servers can be readily discovered and clients communicate with the servers using HTTP/1.1 sessions on TCP connections [3]. We consider a case that each client provides a non-spoofed **ID** (e.g. SYN-cookie [12]), which is utilized to identify the client during our detection period. Owing to the characteristics of this DoS attack, we can assume that the number of attackers $d \ll n$, where n is the total client amount.

2.2.2 Victim/Detection Model

The victim model in the GT-based defense scheme consists of multiple back-end servers, which can be web/application servers, database servers, and distributed file systems. We do not take classic multi-tier web servers as the model, since the detection scheme is deployed directly on the victim tier and identifies the attacks targeting at the same victim tier, thus multi-tier attacks should be separated into several classes to utilize this detection scheme. The victim model along with front-end proxies are shown in Fig. 2.1.

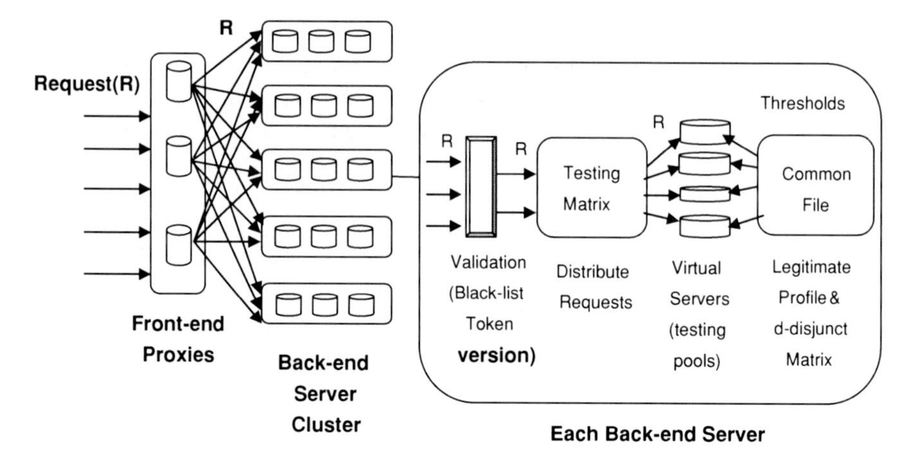

Fig. 2.1 The victim and detection models. This figure also provides an overview of the GT-based detection scheme.

We assume that all the back-end servers provide multiple types of application services to clients using HTTP/1.1 protocol on TCP connections. Each back-end server is assumed to have the same amount of resource. Moreover, the application services to clients are provided by K virtual private servers (K is an input parameter), which are embedded in the physical back-end server machine and operating in parallel. Each virtual server is assigned with equal amounts of static service resources, e.g., CPU, storage, memory, and network bandwidth. The operation of any virtual server will not affect the other virtual servers in the same physical machine. The reasons for utilizing virtual servers are twofold: first, each virtual server can reboot independently, thus is feasible for recovery from possible fatal destruction; second, the state transfer overhead for moving clients among different virtual servers is much smaller than the transfer among physical server machines.

As soon as the client requests arrive at the front-end proxy, they will be distributed to multiple back-end servers for load balancing. On being accepted by one physical server, one request will be simply validated based on the list of all identified attacker **ID**s (black-list). If it passes the authentication, it will be distributed to one virtual server within this machine by means of virtual switch. This distribution depends on the testing matrix generated by the detection algorithms described in Sect. 2.4. By periodically monitoring the average response time to service requests and comparing it with specific thresholds fetched from a legitimate profile, each virtual server is associated with a "negative" or "positive" outcome and a decision over the identities of all clients can be made among all physical servers, as discussed further in Sect. 2.5.

As can be seen, the major components of this model are (1) the Testing Matrix based on which we decide the distribution of clients/requests to virtual servers so that a test can be performed and (2) how to determine the testing results.

2.3 Size Constraint Group Testing

As mentioned in the detection model, each testing pool is mapped to a virtual server within a back-end server machine. Although the maximum number of virtual servers can be extremely huge, since each virtual server requires enough service resources to manage client requests, it is practical to have the virtual server quantity (maximum number of servers) and capacity (maximum number of clients that can be handled in parallel) constrained by two input parameters K and w respectively. Thus the testing matrix must satisfy the two mentioned constraints, thereby extending the traditional GT model as follows:

Definition 2.1 *Size Constraint Group Testing* (*SCGT*): For any binary matrix M, let $w_{\mathrm{i}} = \sum_{j=1}^{n} M[i, j]$ be the weight of the *ith* pool, and t be the number of pools. The model is for identifying d defected items in the minimum period of time, by constructing matrix $M_{\mathrm{t}\times\mathrm{n}}$ and conducting group testing procedures based on that, where $w_i \leq w$ for a given w and $t \leq K$ for a given server quantity K.

The following section presents the solutions to construct such a testing matrix.

2.4 Matrix Construction and Latency Analyses

We present three detection algorithms: sequential detection with packing (SDP), sequential detection without packing (SDoP) and partial non-adaptive detection (PND) in this section. Note that the length of each testing round is a predefined constant P (which is discussed later), hence we analyze the algorithm complexity in terms of the number of testing rounds for simplicity.

Since each client has a unique **ID**, the two terms "**ID**" and "client" are interchangeable in this section. In addition, all the following algorithms are executed in each physical back-end server, which is an independent testing domain as mentioned. Therefore, the term "servers" denote the K virtual servers in this section.

For simplicity, we assume that $n \equiv 0\ (\mathbf{mod}\ K)$, $n > w \geq d \geq 1$ and $|A| \geq d+1$. Notice that the last inequality holds in practice as the properties of application DoS attacks indicate in Sect. 2.1.

2.4.1 Sequential Detection With Packing

This algorithm investigates the benefit of classic sequential group testing, i.e., optimizing the grouping of the subsequent tests by analyzing existing outcomes. Similar to traditional sequential testing, each client (column) only appears in one testing pool (server) at a time. However, to make full use of the available K servers, we have all servers conduct tests in parallel. Details can be found in Algorithm 4.

Algorithm 4 Sequential Detection with Packing (*SDP*)

1: **while** $|S| \neq 0$ **do**
2: **for all** server i in A **do**
3: $w_\text{i} \leftarrow \lceil \frac{|S|}{|A|} \rceil$ // Assign even number of distinct clients to each server.
4: **end for**
5: $G \leftarrow \lceil \frac{n-(|S|-(|A\setminus I|)w_\text{i})}{w} \rceil$ // Identified legitimate **ID**s exempt from the following tests. Their subsequent requests are handled by G non-testing servers, which are selected from the K servers and only provide normal services (no testing).**We call this process as "packing" the clients into non-testing servers.**
6: **for all** server i under attack **do**
7: $Q \leftarrow$ set of **ID**s on i
8: **if** $|Q| = 1$ **then**
9: $S \leftarrow S \setminus Q$ // This **ID** belongs to an attacker. Add it into the black-list.
10: **end if**
11: **end for**
12: $S \leftarrow S \setminus L$ // All **ID**s on safe (with negative test outcome) servers are identified as legitimate.
13: $A \leftarrow$ {server1, ..., server $(K - G)$} // Select the first G servers as non-testing machines.
14: **end while**

2.4.1.1 Algorithm Description

The basic idea of the *SDP* algorithm can be sketched as follows. Given a set S of suspect **ID**s, the algorithm first randomly assigns (i.e. distribute their requests) them to the available testing servers in set A, where each server will receive requests from approximately the same number of clients, roughly $w_\text{i} = \lceil \frac{|S|}{|A|} \rceil$. For each test round, the **ID**s on the negative servers are identified as legitimate clients, and are "packed" into a number G of non-testing machines. Since they need no more tests, only normal services will be provided for the following rounds. As more testing servers will speed up the tests, given at most K server machines in total, G is then supposed to be minimized to the least, as long as all identified legitimate clients can be handled by the non-testing w-capacity servers. Hence $G = \lceil \frac{n-(|S|-(|A\setminus I|)w_i)}{w} \rceil$.

With the assumption $|A| \geq d + 1$, we have at least $|A| - d$ servers with negative outcomes in each testing round, hence at least $(|A| - d)w_\text{i}$ legitimate **ID**s are identified. If any server containing only one active **ID** is found under attack, the only **ID** is definitely an attacker. Then its **ID** is added into the black-list and all its requests are dropped. Iterate the algorithm until all **ID**s are identified as either malicious or legitimate. Via the "packing" strategy, legitimate clients can exempt from the influence of potential attacks as soon as they are identified.

$$
\begin{array}{c} \\ 1\\2\\3\\4\\5 \end{array}
\begin{array}{c}
\begin{array}{cccccccccc} \textcircled{1} & 2 & \textcircled{3} & 4 & 5 & 6 & 7 & 8 & 9 & 10 \end{array} \\
\begin{pmatrix}
1 & 1 & 0 & 0 & 0 & 0 & 0 & 0 & 0 & 0 \\
0 & 0 & 1 & 1 & 0 & 0 & 0 & 0 & 0 & 0 \\
0 & 0 & 0 & 0 & 1 & 1 & 0 & 0 & 0 & 0 \\
0 & 0 & 0 & 0 & 0 & 0 & 1 & 1 & 0 & 0 \\
0 & 0 & 0 & 0 & 0 & 0 & 0 & 0 & 1 & 1
\end{pmatrix}
\end{array}
\overset{testing}{\Longrightarrow}
\begin{pmatrix} 1\\1\\0\\0\\0 \end{pmatrix}
$$

$$
\begin{array}{c} \\ 1\\2\\3\\4\\5 \end{array}
\begin{array}{c}
\begin{array}{cccccccccc} \textcircled{1} & 2 & \textcircled{3} & 4 & 5 & 6 & 7 & 8 & 9 & 10 \end{array} \\
\begin{pmatrix}
1 & 0 & 0 & 0 & 0 & 0 & 0 & 0 & 0 & 0 \\
0 & 0 & 0 & 1 & 0 & 0 & 0 & 0 & 0 & 0 \\
0 & 1 & 1 & 0 & 0 & 0 & 0 & 0 & 0 & 0 \\
0 & 0 & 0 & 0 & 0 & 0 & 1 & 1 & 0 & 0 \\
0 & 0 & 0 & 0 & 1 & 1 & 0 & 0 & 1 & 1
\end{pmatrix}
\end{array}
\overset{testing}{\Longrightarrow}
\begin{pmatrix} 1\\0\\1\\*\\* \end{pmatrix}
$$

$$
\begin{array}{c} \\ 1\\2\\3\\4\\5 \end{array}
\begin{array}{c}
\begin{array}{cccccccccc} \textcircled{1} & 2 & \textcircled{3} & 4 & 5 & 6 & 7 & 8 & 9 & 10 \end{array} \\
\begin{pmatrix}
1 & 0 & 0 & 0 & 0 & 0 & 0 & 0 & 0 & 0 \\
0 & 1 & 0 & 0 & 0 & 0 & 0 & 0 & 0 & 0 \\
0 & 0 & 1 & 0 & 0 & 0 & 0 & 0 & 0 & 0 \\
0 & 0 & 0 & 1 & 0 & 0 & 1 & 1 & 0 & 0 \\
0 & 0 & 0 & 0 & 1 & 1 & 0 & 0 & 1 & 1
\end{pmatrix}
\end{array}
\overset{testing}{\Longrightarrow}
\begin{pmatrix} 1\\0\\1\\*\\* \end{pmatrix}
$$

Fig. 2.2 An example of how SDP algorithm works

Example A 3-round detection example in Fig. 2.2 illustrates the execution of this algorithm. Given $n=10$, $d=2$, $K=5$, and $w=4$. Let $|S| = 10$, $|A| = 5$, and **ID** 1,3 be the attackers. In the first round, $w_i = 2$ yields two **ID**s on each server. Server 1 and server 2 (containing **IDs** 1 and 3 respectively) will indicate being under attack, i.e., $|I| = 2$ (two servers are under attack) and $|A \setminus I| = 3$, pack legitimate **ID**s on the other three servers into $G=2$ servers. Update $|S| = 4$ and $|A| = 3$. In the second round, $w_i = 2$ yields at most two **ID**s on each server, assume that server 1 and server 3 are under attack, then again pack the legitimate **ID** 4 in server 2 to server 4 (server 5 is already full). In the third round, servers 1 and 3 only contain attacker **IDs** 1 and 3, respectively. Therefore, the two attackers are identified at the end of this round. What we would like to point out is, in this instance, for every new assignment, no testing server contains multiple malicious **ID**, which will affect more servers and yield to the least number of safe servers at each round, and thus is likely to cost the most testing rounds. Hence, this is an instance of worst performance, and its result verifies our $O(\log_{\frac{K'}{d}} \frac{n}{d})$ upper bound mentioned in the below analysis.

2.4.1.2 Performance Analysis

With regard to the computational overhead of the matrix M in this context, it is a re-mapping of the suspect clients to the testing servers based on simple strategies and previous testing outcomes, thus is negligible and up to $O(1)$. The time cost of each testing round is at most P, including the recomputation time of the testing matrix.

Besides this, the overall time complexity of this algorithm in terms of the number of testing rounds are depicted in the following theorems.

Lemma 2.1 The number of available testing server machines is $|A| \in \left[K - \lceil \frac{n-d}{w} \rceil, K\right]$.

proof In each round, the total number of legitimate **ID**s $n - (|S| - (|A \setminus I|)w_i)$ is non-decreasing. Hence, the number of servers used for serving identified legitimate **ID**s, $G = \lceil \frac{n-(|S|-(|A \setminus I|)w_i)}{w} \rceil$ is non-decreasing. Therefore $|A| = K - G$ will finally converge to $K - \lceil \frac{n-d}{w} \rceil$. □

Theorem 2.1 *The SDP algorithm can identify all the d attackers within at most* $O(\log_{\frac{K'}{d}} \frac{n}{d})$ *testing rounds, where* $K' = K - \lceil \frac{n-d}{w} \rceil$.

proof In each round, at most d servers are under attack, i.e., at most $\frac{d|S|}{|A|}$ **ID**s remain suspect. Since at the end of the program, all legitimate **ID**s are identified, at most d malicious **ID**s remain suspect. If they happen to be handled by d different servers, i.e., each positive server contains exactly one attacker, then the detection is completed. If not, one more testing round is needed. Assume that we need at most j testing rounds in total, then $n\left(\frac{d}{|A|}\right)^{j-1} = d$ where $K - \lceil \frac{n-d}{w} \rceil \le |A| \le K$. Therefore, $j \le \log_{(\frac{K'}{d})} \frac{n}{d} + 1$, where $K' = K - \lceil \frac{n-d}{w} \rceil$. □

Remark Note that the packing assignment in the SDP algorithm reflects the following design principle: serve all legitimate clients with the least number of good servers (non-testing machines) in order to achieve the maximum number of testing servers, thereby minimizing the testing rounds. In addition, it also ensures that all legitimate requests can be directed to a safe server as soon as possible.

2.4.2 Sequential Detection Without Packing

Considering the potential overload problem arises from the "packing" scheme adopted in *SDP*, we present the SDoP algorithm of which legitimate clients do not shift to other servers after they are identified. This emerges from the observation that legitimate clients cannot affect the test outcomes since they are negative. Algorithm 5 includes the abstract pseudocode of this *SDoP* scheme.

Algorithm 5 Sequential Detection without Packing (*SDoP*)

1: $w_i \leftarrow$ number of **ID**s on server i;
2: **for all** server i **do**
3: $\quad w_i \leftarrow \lceil \frac{S}{A} \rceil$ // Evenly assign clients to servers as *SDP* did.
4: **end for**
5:
6: **while** $|S| \neq 0$ **do**

7: Randomly reassign $|S|$ suspect **ID**s to K servers, and keep legitimate **ID**s unmoved.
8: $L \leftarrow$ set of **ID**s on safe servers
9: $S \leftarrow S \setminus (S \cap L)$ // $|S \cap L|$ **ID**s are identified as legitimate.
10: **for all** server i under attack **do**
11: $Q \leftarrow$ set of **ID**s on i
12: **if** $|Q \cap S| = 1$ **then**
13: $S \leftarrow S \setminus Q$// The clients in $Q \cap S$ are attacker and added into the black-list.
14: **end if**
15: **end for**
16: **for all** servers i **do**
17: **if** $w_i = w$// The volume approaches the capacity. **then**
18: Reassign all $n - |S|$ legitimate **ID**s to K servers, and go to 6 // Load balancing.
19: **end if**
20: **end for**
21: **end while**

2.4.2.1 Algorithm Description

The basic idea of the *SDoP* algorithm can be sketched below. Given suspect **ID**s set S with an initial size n, evenly assign them to the K server machines, similar to *SDP* in the first round. For the following rounds, assign suspect **ID**s to the K servers instead of $|A|$ available ones. For the identified legitimate IDs, never move them until their servers are to be overloaded. In this case, reassign all legitimate **ID**s over the K machines to balance the load. For a server with the positive outcome, the **ID**s active on this server but not included by the set of identified legitimate ones are still be identified as suspect. However, if there is only one suspect **ID** of this kind in a positive server, this **ID** is certainly an attacker.

Remarks Compared with SDP, if no overloading happens, the SDoP algorithm avoids moving identified legitimate clients, thereby decreasing the reassignment expense and increasing the number of available testing machines. However, if w is relatively small, the algorithm also has to reassign all legitimate clients to K servers and balance the weight w_i on different servers. Once overloading happens, the reason for reassigning both legitimate and suspect ones into K servers at two different testing rounds, instead of reassigning them together at one time, is to avoid the possibility that all non-testing (safe) servers after the reassignment contain no suspect client and much more legitimate clients than others, i.e, a large w_i. We will analyze the performance of this algorithm in the following section with regard to different w scope.

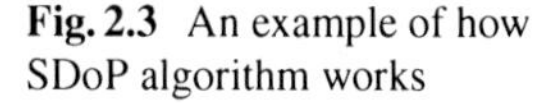

Fig. 2.3 An example of how SDoP algorithm works

$$
\begin{array}{c|cccccccccc|c}
 & \textcircled{1} & 2 & \textcircled{3} & 4 & 5 & 6 & 7 & 8 & 9 & 10 & \\
1 & 1 & 1 & 0 & 0 & 0 & 0 & 0 & 0 & 0 & 0 & 1 \\
2 & 0 & 0 & 1 & 1 & 0 & 0 & 0 & 0 & 0 & 0 & 1 \\
3 & 0 & 0 & 0 & 0 & 1 & 1 & 0 & 0 & 0 & 0 & 0 \\
4 & 0 & 0 & 0 & 0 & 0 & 0 & 1 & 1 & 0 & 0 & 0 \\
5 & 0 & 0 & 0 & 0 & 0 & 0 & 0 & 0 & 1 & 1 & 0
\end{array}
\quad \overset{testing}{\Longrightarrow}
$$

$$
\begin{array}{c|cccccccccc|c}
 & \textcircled{1} & 2 & \textcircled{3} & 4 & 5 & 6 & 7 & 8 & 9 & 10 & \\
1 & 0 & 0 & 1 & 0 & 0 & 0 & 0 & 0 & 0 & 0 & 1 \\
2 & 0 & 1 & 0 & 0 & 0 & 0 & 0 & 0 & 0 & 0 & 0 \\
3 & 0 & 0 & 0 & 1 & 1 & 1 & 0 & 0 & 0 & 0 & 0 \\
4 & 1 & 0 & 0 & 0 & 0 & 0 & 1 & 1 & 0 & 0 & 1 \\
5 & 0 & 0 & 0 & 0 & 0 & 0 & 0 & 0 & 1 & 1 & 0
\end{array}
\quad \overset{testing}{\Longrightarrow}
$$

Example Figure 2.3 contains a 2-round detection example to illustrate the execution of this algorithm. Assume $n=10$, $d=2$, $K=5$ and the client 1, 3 are the attackers. In the first testing round, $w_{\mathrm{i}} = 2$ yields two **ID**s on each server, then only servers 1 and 2 will have positive outcomes. Update suspect **ID** set $S = \{1, 2, 3, 4\}$. In the second testing round, since no attackers have been identified and no servers are overloaded, we reassign all suspect **ID**s to K servers while not moving the legitimate clients, and servers 1 and 4 turn out to be under attack. Hence the legitimate **ID** set is $L = \{2, 4, 5, 6, 7, 8, 9, 10\}$ and the two attackers are captured within two testing rounds.

2.4.2.2 Performance Analysis

It is trivial to see that the computational overhead of the testing matrix M is similar to that of the SDP algorithm, therefore it can be ignored. The following theorems exhibit the overall detection delay of the algorithm, in terms of the number of testing rounds.

Lemma 2.2 *The number of* **ID***s w_{i} on server i does not exceed server capacity w till round j, where $j = \log_{\frac{K}{d}} \frac{n}{n-w(K-d)}$.*

proof At most d servers get positive outcomes in each round. If we assume round j is the last round that has no overloaded servers, then the maximum number of **ID**s on one server at round j is: $w^{\mathrm{j}}_{\max} = \sum_{i=1}^{j} (\frac{n}{K})(\frac{d}{K})^{\mathrm{i}-1}$. Since round j+1 will have at least one overloaded server, we have $w^{\mathrm{j}}_{\max} = w$, then $j = \log_{\frac{K}{d}} \frac{n}{n-w(K-d)}$. □

Lemma 2.3 *All malicious* **ID***s are identified within at most $O(\log_{\frac{K}{d}} \frac{n}{d})$ testing rounds if $w \geq \frac{n-d}{K-d}$.*

proof In each round with no overloaded servers, at most $\frac{d}{K}|S|$ legitimate **ID**s remain suspect. After $i = \log_{\frac{K}{d}} \frac{n}{K}$ rounds at most K suspect **ID**s are left, so we need only one more round to finish the testing. Hence, we need $i + 1 = \log_{\frac{K}{d}} \frac{n}{d}$ rounds. Since it requires that no servers are overloaded within these rounds, we have $i + 1 < j = \log_{\frac{K}{d}} \frac{n}{n-w(K-d)}$ which yields $w \geq \frac{n-d}{K-d}$. □

Lemma 2.4 *All malicious* **ID***s are identified within at most* $O(\log_{\frac{K}{d}} \frac{n}{d})$ *testing rounds if* $1 \leq w \leq \frac{n-d}{K-d}$.

proof If $w \leq \frac{n-d}{K-d}$, reassignments for legitimate **ID**s will be needed after round $j = \log_{\frac{K}{d}} \frac{n}{n-w(k-d)}$. Assume that the algorithm needs x more testing rounds in addition to $i + 1 = \log_{\frac{K}{d}} \frac{n}{d}$ to reassign and balance legitimate **ID**s on all servers, since the position changes of these legitimate **ID**s will neither influence test outcomes nor delay the identifications of suspect set S (they take place simultaneously), we hence have

$$
\begin{aligned}
x &\leq i + 1 - j \\
&\leq \log_{\frac{K}{d}} \frac{n}{d} - \log_{\frac{K}{d}} \frac{n}{n - w(k - d)} \\
&\leq \log_{\frac{K}{d}} \frac{n - w(K - d)}{d} \leq \log_{\frac{K}{d}} \frac{n}{d}
\end{aligned}
$$

Therefore, the total number of testing rounds is at most $O(\log_{\frac{K}{d}} \frac{n}{d})$. □

Theorem 2.2 *The* SDoP *algorithm can identify all attackers within at most* $O(\log_{\frac{K}{d}} \frac{n}{d})$ *testing rounds.*

Proof Directly from Lemmas 2.3 and 2.4. □

2.4.3 Partial Non-Adaptive Detection

Considering the fact that in the two sequential algorithms mentioned, we could not identify any attackers until we isolate each of them to a virtual server with a negative outcome, which may increase the detection latency. Therefore, we present a hybrid of sequential and non-adaptive method in this section. In this scenario, the requests from the same client will be received and responded by different servers in a round-robin manner. Different from *SDP* and *SDoP*, a d-disjunct matrix is used as the testing matrix in this scheme and attackers can be identified without the need of isolating them into servers.

As mentioned in Chap. 1, matrix M is called d-disjunct if no single column is contained in the boolean sum of any other d columns, and all positive items can be identified within one round. Therefore we adopt this method as a subroutine to generate M for each testing round. For the following analysis, we refer to this algorithm as *dDisjunct* (n,d) (Algorithm 2), which takes n,d as input, and outputs a d-disjunct matrix with n columns.

We now introduce the *PND* detection method using d-disjunct matrix with an assumption that the row weight of the constructed matrix does not exceed the server capacity w. Let $t(d,n)$ denote the number of rows required in the d-disjunct matrix with n columns obtained from Algorithm 2. We need to consider different ranges of input K with respect to $t(d,n)$ as follows:

Case 1 $K \geq t(d,n)$. In this case, all n suspect **ID**s can be assigned to $t(d,n)$ servers according to a d-disjunct matrix, and thus we can identify all malicious IDs with one testing round.

Case 2 $K < t(d,n)$. With inadequate servers to test all n clients in parallel, we can iteratively test a part of them, say $n' < n$, where $t(d,n') \leq |A|$. The detailed scheme is depicted as follows and the pseudocode is shown in Algorithm 6.

2.4.3.1 Algorithm Description

In the PND algorithm, all n suspect **ID**s are evenly distributed to all K servers as the other algorithms did at the beginning. After the first round, all identified legitimate **ID**s are packed into $G = \lceil \frac{n-|S|}{w} \rceil$ servers. Therefore, $|A| = K - G$ servers are available for testing afterwards. If $|A| \geq t(d,|S|)$, we can construct a d-disjunct matrix $M_{|A| \times |S|}$ by calling function $dDisjunct(|S|, d)$ and identify all **ID**s in one more testing round. Otherwise, if $|A| < t(d,|S|)$, we can first find the maximum n' $(1 \leq n' \leq |S|)$ such that $t(d,n') \leq |A|$; secondly, partition set S into two disjoint sets S_1 and S_2 with $|S_1| = n'$; thirdly pack all S_2 to G non-testing server machines and call subroutine $dDisjunct(|S_1|, d)$ with updated d to identify all **ID**s in S_1; and finally drop the attackers and pack all legitimate **ID**s into G machines and continue to test $S = S_2$. Iterate this process until all **ID**s are identified.

2.4.3.2 Performance Analysis

The time complexity of the function *dDisjunct*(n,d) (Algorithm 2) is $O(sqn)$, i.e. $O(Kn)$ with $sq \leq K$, which can be decreased to $O(1)$ time cost in real-time. Since it is too complicated to use the upper bound in Theorem 1.5 for $t(d,n)$, in this analysis, we instead use a simple bound $t(d,n) = n$ to obtain a rough estimate on the algorithm complexity in terms of the number of rounds. Moreover, we further investigate its performance based on an optimal d-disjunct, so as to demonstrate the potential of this scheme.

Algorithm 6 Partial Non-adaptive Detection with $K < t(d,n)$

1: Evenly assign n **ID**s to K servers with no two servers having the same client IDs. Pack all legitimate **ID**s on into G servers.
2: $|A| \leftarrow K - G$ // Update the number of testing servers.

3: **if** $|A| \geq t(d, |S|)$ **then**
4: Run $dDisjunct(d, |S|)$ (Algorithm 2)
5: decode from the obtained d-disjunct matrix and identify all IDs;
6: $S \leftarrow 0$; // Testing finishes.
7: **else**
8: **while** $|S| \neq 0$ **do**
9: Partition S into S_1, S_2 where $|S_1| = n'$ and $n' = \max n'' \in [1, |S|]$ satisfying $t(d,n'') \leq |A|$.
10: Run $dDisjunct(d, |S_1|)$, decode from the obtained d-disjunct matrix and identify all **ID**s in S_1, update d.
11: Pack legitimate **ID**s in S_1 to G machines, update G and A; $S \leftarrow S_2$.// Iteration with suspect **ID** set S_2.
12: **end while**
13: **end if**

Lemma 2.5 *The* PND *algorithm identifies all the attackers within at most* $O(dw^2/(Kw - w - n))$ *testing rounds.*

Proof Let us consider the worst case. In each round, assuming no attackers are identified and filtered out previously, we have $t(d, n') + \lceil \frac{n-n'}{w} \rceil = K$. Since $t(d, n') \leq n'$ then $n' \geq \frac{Kw-w-n}{w-1}$. Therefore the maximum number of testing rounds needed is: $\frac{dn}{K} / \frac{Kw-w-n}{w-1} \leq dw^2/(Kw - w - n) = O(dw^2/(Kw - w - n))$. □

This complexity for PND is not as good as that of the previous two algorithms, partially due to the simple bound used. Therefore, we continue the analysis using the tight lower bound for $t(d,n)$ proposed in [13] and investigate the corresponding performance of PND as follows.

Lemma 2.6 *(D'yachkov-Rykov lower bound)* [13] *Let* $n > w \geq d \geq 2$ *and* $t > 1$ *be integers. For any superimposed* $(d-1, n, w)$*-code* ((d, n, w)*-design*) X *of length* t (X *is called a* $(d-1)$*-disjunct matrix with* t *rows and* n *columns), the following inequality holds*:

$$t \geq \lceil \frac{dn}{w} \rceil$$

The following are the related performance analysis based on this lower bound of $t(d,n)$.

Corollary 2.1 *In the* PND *algorithm, given* $w \geq d \geq 1$, *we have:* $n' = \min\{\frac{|A|w}{d+1}, n - Kw + w + |A|w\}$

proof According to Lemma 2.6, with number of columns $n'' \in [1, |S|]$, we have

$$|A| \geq \lceil \frac{(d+1)n''}{w} \rceil \geq \frac{(d+1)n''}{w} \Rightarrow n' = \max\, n'' \leq \frac{|A|w}{d+1}$$

Meanwhile, in the *PND* algorithm, for each round i, we have:

$$|A_i| \geq K - \lceil \frac{n-n'}{w} \rceil \Rightarrow n' \leq n - Kw + w + |A|w$$

□

Lemma 2.7 *In any testing round i*:

1. $n' = \frac{|A_i|w}{d+1}$ when $K \in \left(d, \frac{dw+n+w}{w}\right]$;
2. $n' = n - Kw + w + |A_i|w$ when $K \in \left[k_1, \frac{dn+n+w}{w}\right)$ with

$$k_1 = \frac{dw + w + n + \sqrt{(dw+w+n)^2 + 4wnd^2}}{2w}$$

Proof According to Corollary 2.1, we have:

1. We have

$$K \leq \frac{dw+n+w}{w} \Leftrightarrow wK \leq dw + n + w$$

$$|A| \geq d + 1 \Leftrightarrow \frac{|A|wd}{d+1} \geq wd$$

   Hence

$$n + w - Kw + |A|w \geq \frac{|A|w}{d+1}$$

2. In order to get

$$\frac{|A|w}{d+1} \geq n - Kw + w + |A|w$$

   we need

$$|A| \leq \frac{((K-1)m - n)(d+1)}{wd}$$

   which requires

$$K - \frac{(K-d)n}{Kw} \leq \frac{((K-1)m - n)(d+1)}{wd}$$

   Hence we need

$$wK^2 - (dw + w + n)K - nd^2 \geq 0$$

   Solving this inequality, we have $K \in [k_1, +\infty)$. Note if $K \geq \lceil \frac{dn+n}{w} \rceil$, we need not do partitions in *PND* algorithm and since

$$k_1 \leq \frac{dn + n + w}{w}$$

we have

$$K \in [k_1, \frac{dn + n + w}{w})$$

Moreover,

$$k_1 > \frac{dw + w + n}{w} \geq \frac{d^2 + w + n}{w}$$

there are thus no overlaps between these two intervals. □

Therefore, we split $K \in (d, +\infty)$ into four disjoint intervals and study which interval of value K yields $O(1)$ testing rounds in worst case besides the interval $K \in [\frac{dn+n+w}{w}, +\infty)$ shown above, as well as complexity for other intervals.

- **I**: $K \in (d, \frac{dw+n+w}{w}]$ yields $n' = \frac{|A|w}{d+1}$;
- **II**: $K \in (\frac{dw+n+w}{w}, k_1)$ yields $n' = \min\{\frac{|A|w}{d+1}, n - Kw + w + |A|w\}$;
- **III**: $K \in [k_1, \frac{dn+n+w}{w})$ yields $n' = n - Kw + w + |A|w$;
- **IV**: $K \in [\frac{dn+n+w}{w}, +\infty)$ yields ONE testing round in total.

Lemma 2.8 *The* PND *algorithm needs at most* $O(1)$ *testing rounds with* $K \in [k_2, \frac{dw+n+w}{w}]$, *where*

$$d \leq \frac{w + \sqrt{w^2 - 4n^2(n - w)}}{2(n - w)}$$

and

$$k_2 = \frac{n + w + \sqrt{n^2 + w^2 + 2nw - 4n^2w + 4d^2wn + 4dwn}}{2w}$$

Proof Since at least one server gets positive outcome at the first testing round, we have

$$|A_0| \geq K - \lceil\frac{(K - 1)n}{Kw}\rceil$$

With simple algebraic computations, we can reach the interval $\left[k_2, \frac{dw+n+w}{w}\right]$ on the condition that

$$d \leq \frac{w + \sqrt{w^2 - 4n^2(n - w)}}{2(n - w)}$$

within interval **I**; however, for intervals **II** and **III**, no such detailed subintervals of K yielding $O(1)$ testing rounds can be obtained. □

Lemma 2.9 *Within interval* **I**, PND *algorithm can identify all* **ID***s with* $O(d + \frac{K}{\sqrt{n}})$ *testing rounds.*

Proof We derive the time complexity from the following recurrence:
Starting round 0: $|S_0| \leq \frac{dn}{K}$, and $K - \lceil \frac{(K-1)n}{Kw} \rceil \leq |A_0| \leq K - \lceil \frac{(K-d)n}{Kw} \rceil$
Ending round T: $0 < |S_T| \leq \frac{|A_T|w}{d+1}$
Iteration: For $\forall i \in [0, T-1]$ we have

$$|S_{i+1}| = |S_i| - \frac{|A_i|w}{d+1}$$

and

$$K - \lceil \frac{n - |S_{i+1}|}{w} \rceil \leq |A_{i+1}| \leq K - \lceil \frac{n - |S_{i+1}| - d}{w} \rceil$$

hence

$$\begin{cases} |A_{i+1}| \geq K - \frac{n}{w} + \frac{|S_i|}{w} - \frac{|A_i|}{d+1} - 1 \\ S_0 \leq \frac{\sum_{i=0}^{T} |A_i| w}{d+1} \end{cases}$$

In order to estimate the maximum time cost, use $|S_0| = \frac{dn}{K}$ to initiate the worst starting case. Solving this recurrence, we get the following inequality:
$\frac{Kw}{2(d+1)} T^2 - (\frac{Kw}{2(d+1)} + \frac{dn}{K} - K + \frac{n}{w} + w)T - (K - \frac{(K-1)n}{w} - \frac{dn(d+2)}{K} + w - 1) \leq 0$, therefore

$$T \leq \frac{d+1}{Kw}\left(\alpha + \sqrt{\frac{\beta}{d+1} + \alpha^2}\right)$$

where

$$\alpha = \frac{Kw}{2(d+1)} + \frac{dn}{K} + \frac{n}{w} - K + w$$

and

$$\beta = 2K^2w - 2K^2n + 2Kn - 2Kw + 2Kw^2 - 2d(d+2)nw$$

Since $\frac{n}{w} \leq K$, $wK \leq dw + n + w$ and $n > w$, we have $\alpha > 0$ and $\beta > 0$. So with trivial computation we can get

$$\begin{aligned} T &\leq \frac{2\alpha(d+1)}{Kw} + \sqrt{\frac{\beta}{d+1}} \\ &< 1 + \frac{2(d+1)^2}{K} + \sqrt{\frac{(4K-2)(d+1)}{n} + \frac{2(d+1)}{d+1}} \\ &< 1 + 2(d+1) + \sqrt{\frac{4K(d+1)}{n} + 2} \\ &< 3 + \sqrt{2} + 2d + \frac{2K}{\sqrt{n}} \end{aligned}$$

Therefore, *PND* will complete the identification within at most $O(d + \frac{K}{\sqrt{n}})$ testing rounds. □

Note that since K is always much smaller than n, the complexity will approach to O(d) in fact.

Lemma 2.10 *Within interval* **III**, *PND can identify all* **ID**s *with at most O(d) testing rounds.*

Proof Similarly, we can get $T \leq 2d+1+2\sqrt{\frac{1+d}{4}}$ by solving the following recurrence
Starting round 0: $|S_0| = \frac{dn}{K}$ and $K - \lceil\frac{(K-1)n}{Kw}\rceil \leq |A_0| \leq K - \lceil\frac{(K-d)n}{Kw}\rceil$
Ending round T: $0 < |S_T| \leq n - Kw + w + |A_T|w$
Iteration: $\forall i \in [0, T-1]$, $|S_{i+1}| = |S_i| - (n - Kw + w + |A_i|w)$ and $K - \lceil\frac{n-|S_{i+1}|}{w}\rceil \leq |A_{i+1}| \leq K - \lceil\frac{n-|S_{i+1}|-d}{w}\rceil$. Hence, *PND* will complete the identification within at most O(d) testing rounds. □

Corollary 2.2 *Within interval* **II**, PND *algorithm can identify all* **ID**s *with* $O(d+\frac{K}{\sqrt{n}})$ *testing rounds.*

Proof According to Lemmas 2.9 and 2.10, the time complexity of *PND* algorithm depends on the value of *n'* at each round, and since *n'* within interval **II** oscillates between $\frac{|A|w}{d+1}$ and $n - Kw + w + |A|w$, the time complexity is at most $O(d + \frac{K}{\sqrt{n}})$. □

Theorem 2.3 *Given* $1 \leq d \leq w$, *the PND algorithm can identify all* **ID**s *within*

1. *at most* $O(d + \frac{K}{\sqrt{n}})$ *testing rounds when* $K \in (d, k_1)$, *whilst at most* $O(1)$*testing rounds when* $K \in [k_2, \frac{dw+n+w}{w}]$ *on condition that* $d \leq \frac{w+\sqrt{w^2-4n^2(n-w)}}{2(n-w)}$;
2. *at most* $O(d)$ *testing rounds when* $K \in [k_1, \frac{dn+n+w}{w})$;
3. *at most* $O(1)$ *testing rounds when* $K \in [\frac{dn+n+w}{w}, +\infty)$; *where*
   $k_1 = \frac{dw+w+n+\sqrt{(dw+w+n)^2+4wnd^2}}{2w}$
   and $k_2 = \frac{n+w+\sqrt{n^2+w^2+2nw-4n^2w+4d^2wn+4dwn}}{2w}$.

Despite the number of needed testing rounds differing for the three algorithms above, the time complexity of calculating each testing round for these algorithms is approximate in practice. It is trivial to see that the costs for *SDP* and *SDoP* are negligible, but not for *PND* algorithm which involves polynomial computation on Galois Field. However, considering that the upper bound of both the number of clients n and attackers d are estimated, the detection system can precompute the d-disjunct matrices for all possible (n,d) pairs offline, and fetch the results in real-time. Therefore, the overhead can be decreased to $O(1)$ and the client requests can be smoothly distributed at the turn of testing rounds without suffering from long delays of matrix update. In some cases, if an online construction of M is required, we can use a randomized construction method which is presented in Chap. 4. This method has a very low computational overhead.

2.5 Detection System Configuration

In this section, we present the rest of the system configuration which illustrates how to implement the above mathematical framework into a practical solution for an application DoS defense . More specifically, we discuss *how to distribute client requests based on M with a low overhead* and *how to generate test outcome with high accuracy.*

2.5.1 System Overview

As mentioned in the GT-based detection model, each back-end server works as an independent testing domain, where all virtual servers within it serve as testing pools. In the following sections, we only discuss the operations within one back-end server, and it is similar in all other servers. The detection consists of multiple testing rounds, and each round can be sketched in four stages (Fig. 2.4):

1. **Stage 1**: Matrix M for testing is generated and updated.
2. **Stage 2**: All clients are distributed to virtual servers based on M. The back-end server maps each client into one distinct column in M and distributes an encrypted token queue to it. Each token in the token queue corresponds to a 1-entry in the mapped column. i.e., client j receives a token with destination virtual server i *iff* $M[i, j] = 1$. Being piggybacked with one token, each request is forwarded to a virtual server by the virtual switch. In addition, requests are validated on arriving at the physical servers for faked tokens or identified malice **ID**. This procedure ensures that all the client requests are distributed exactly as how the matrix M regulates, and prevents any attackers from accessing the virtual servers other than the ones assigned to them.
3. **Stage 3**: All the servers are monitored for their service resource usage periodically. Specifically, the arriving request aggregate (the total number of incoming requests) and average response time of each virtual server are recorded and compared with some dynamic thresholds (to be shown later). All virtual servers are associated with positive or negative outcomes accordingly.
4. **Stage 4**: Based on these testing outcomes and M, decode and identify legitimate or malicious **ID**s. By following the detection algorithms discussed above, all the attackers can be identified within several testing rounds.

To lower the overhead and delay introduced by the mapping and piggybacking for each request, the system is exempted from this procedure in normal service state. As shown in Fig. 2.5, the back-end server cycles between two states, which we refer to as NORMAL mode and DANGER mode. Once the estimated response time (ERT) of any virtual server exceeds some profile-based threshold, the whole back-end server will transfer to the DANGER mode and execute the detection scheme. Whenever the

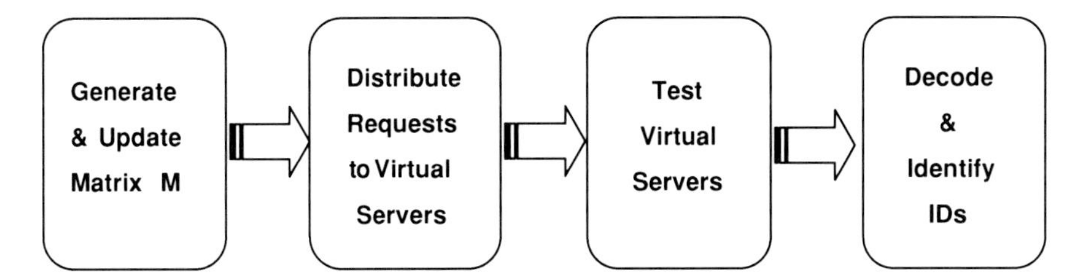

Fig. 2.4 The four stages within one testing round in DANGER mode

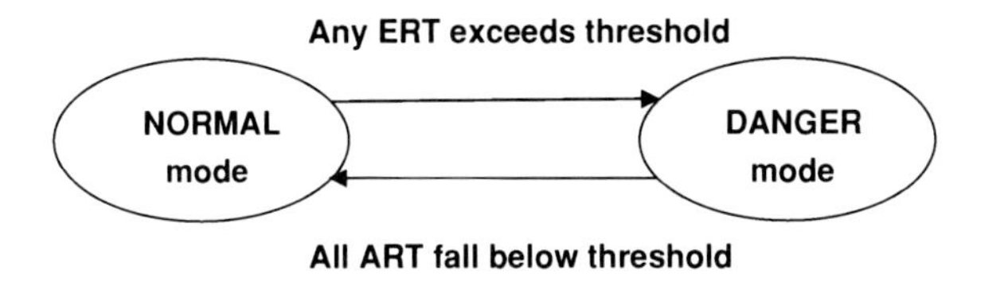

Fig. 2.5 Two-state diagram of the system. Once the estimated response time (ERT) of any virtual server exceeds some profile-based threshold, the whole back-end server will transfer to the DANGER mode and execute the detection scheme. Whenever the average response time (ART) of each virtual server falls below the threshold, the physical server returns to NORMAL mode

average response time (ART) of each virtual server falls below the threshold, the physical server returns to NORMAL mode.

2.5.2 Configuration Details

Distributing Tokens. Two main purposes of utilizing tokens are associating each client with a unique, non-spoofed **ID** and assigning them to a set of virtual servers based on the testing matrix. On receiving the connection request from a client, each back-end server responses with a token queue where each token is of 4-tuple: (client ID, virtual server ID, matrix version, encrypted key). "client ID" refers to the unique non-spoofed **ID** for each client, which we assume unchanged during the testing period (DANGER mode). "virtual server ID" is the index of each virtual server within the back-end server. This can be implemented as a simply index value, or through a mapping from the IP addresses of all virtual servers. The back-end server blocks out-of-date tokens by checking their "matrix version" value, to avoid messing up the request distribution with non-uniform matrices. With regard to the "encrypted key", it is an encrypted value generated by hashing the former three values and a secured service key. This helps rule out any faked tokens generated by attackers.

Length of Testing Round. Since we need to distribute the requests exactly the way M regulates, it is possible that: if P is too short, some clients may not have distributed their requests to all their assigned servers, i.e., not all the 1-entries in M are matching with at least one request. For an attacker who launches a low-rate high-workload attack, its high-workload requests may only enter a part of its assigned servers, so false negative occurs in this case. However, if P is too long, the detection latency will be significantly increased. Therefore, P must be carefully decided to be just long enough for each client to spread their requests to all the assigned servers, i.e., each column needs to be mapped with at least $\sum_{i=1}^{t} M[i, j]$ requests. Thus, we have:

$$P = \max_{j=1}^{n} \sum_{i=1}^{t} M[i, j]/r_{\min}$$

where $r_{\min}$ denotes the minimum interarrival request rate, provides a theoretical lower bound of P.

Legitimate Profile. The legitimate profile does not refer to the session behavior profile for legitimate clients, but instead records the distribution of the *ART* on a virtual server receiving only legitimate traffic. Malicious requests are certainly to generate destructions to the victim server machines, whose *ART* will usually be much higher than that of normal cases. Therefore, *ART* can work as an indicator of the application resource usage. However, the resource usage varies for different time intervals (peak/non-peak time) due to the change of client quantity, so we also investigate the *ART* distributions regarding each possible number of clients, assuming that there are at most n clients.

A sample construction of this profile is: The distributions of *ART* in legitimate traffic at different time intervals for several weeks are obtained after the system is established. Legitimate traffic can be achieved by de-noising measures and ruling out the influences of potential attacks [14]. The *ith* entry of the profile records the distribution of *ART* on a virtual server with $i \in [1, n]$ clients. Specifically, we assign i clients to a virtual server, and evenly divide the round length P into 100 subintervals (which we refer as P_{sub} throughout the paper), and monitor the server *ART* within each subinterval P_{sub}.

NORMAL mode and Transfer Threshold. To decrease the length of detection period, where additional state maintenance overhead and service delay cannot be avoided, the back-end server cluster provides normal service to clients, without any special regulations. This is referred as NORMAL mode, which takes *ERT* (estimated response time) as a monitoring object. Note that *ERT* is an expected value for *ART* in the near future, and can be computed as:

$$ERT = (1 - \alpha) \cdot ERT + \alpha \cdot ART$$

where α is decay weight for smoothing fluctuations. Since the interarrival rate and workload of client request can be randomized distributed, it is difficult to perfectly fit the *ART* distribution using classic distribution functions. Considering that Normal Distribution [15] can provide an approximate fitting to *ART*, a simplified threshold can be adpoted based on the Empirical Rule as follows: If any virtual server has an $ERT > \mu + 4\sigma$ (μ and σ denote the expected value and standard deviation of the fitted *ART* distribution), the back-end server is probably undergoing an attack, and thus transfers to DANGER mode for detection. To enhance the accuracy of this threshold, more sophisticated distributions can be employed for fitting the samples, however, higher computation overhead will be introduced for the threshold.

DANGER mode and Attack Threshold. In DANGER mode, besides the *ART* values, the back-end server simultaneously counts the arriving request aggregate within each P_{sub} for each virtual server. The motivation of this strategy arises from the fact that, high-rate DoS attacks always saturate the server buffer with a flood of malicious requests. By counting the number of arrivings in the buffer requests periodically, possible high-rate attacks can be detected even before depleting the service resources.

We predefine R as a maximum legitimate interarrival rate, and derive an upper bound of the arriving request aggregate C_{j} for virtual server j within period P_{sub} as:

$$C_{\mathrm{j}} = \sum M[j,i] \cdot \lfloor \frac{\sum_{s=j+1}^{t} M[s,i] + R \cdot P_{\mathrm{sub}}}{\sum_{k=1}^{t} M[k,i]} \rfloor$$

Once this threshold is violated in a virtual server, it undergoes high-rate attacks or flash-crowd traffic. A positive outcome can be generated for this testing round.

The outcome generation by monitoring the *ART* value for a virtual server consists of the following detailed steps:

1. Check the *ART* distribution profile for the values of parameters μ and σ, as mentioned in the NORMAL mode section;
2. among all the 100 subintervals P_{sub} (each is $P/100$) within the current testing period P, if no violation: $ART \geq \mu + 4\sigma$ occurs in any P_{sub}, the virtual server gets negative outcome for this testing round;
3. if for some subinterval P_{sub}s, $ART \geq \mu + 4\sigma$ occurs, this virtual server is in danger, either under attack, or undergoing flash-crowd traffic. In this case, we wait till the end of this round to get the distribution of *ART* values for all P_{sub} s for further decision;
4. if the ratio of "danger" subintervals P_{sub} s with $ART \geq \mu + 4\sigma$, over the total subinterval amount (100 in this case), exceeds some quantile regulated by the *Empirical Rule*, e.g., quantile 4% for confidence interval $[\mu + 2\sigma, \infty)$, this virtual server will be labeled as positive;
5. for the other cases, the virtual server will have a negative outcome.

After identifying up to d attackers, the system remains in the DANGER mode and continues monitoring the *ART* for each virtual server for one more round. If all the virtual servers have negative outcomes, this back-end server is diagnosed as healthy and return to the NORMAL mode, otherwise further detections will be executed (because of error tests).

With regard to the two objects monitored, C_{j} provides different counting thresholds for different virtual servers, while *ART* profile supplies dynamic thresholds for virtual server containing different amounts of client. The combination of these two dynamic thresholds helps decrease the testing latency and false rate.

2.6 Experimental Analysis

To demonstrate the theoretical complexity results shown in the previous section, we present here a simulation study which has been conducted by [16] in terms of four metrics: *average testing latency* T which refers to the length of the time interval from attackers starting sending requests till all of them are identified; *average false positive rate* f_p and *false negative rate* f_n; as well as the *average number of testing rounds* R_{test} which stands for the number of testing rounds needed for identifying all the clients by each algorithm.

2.6.1 Configurations

To this end, a simulator in Java by modeling both the n clients and K virtual servers as independent threads was implemented. In order to mimic the real client (including attacker) behavior and dynamic network environment, the client/server system was implemented as follows:

- each legitimate client joins in and leaves the system at random times which are uniformly distributed, while the attacker threads arrive at time $t=30$ s and keep live until being filtered out.
- both legitimate and malicious clients send requests with a random interarrival rate and CPU processing time (workload) to the virtual servers, however, legitimate ones have a much smaller random range than that of the attackers.
- each virtual server is equipped with an infinite request buffer and all the client requests arrive at the buffers with 0 transmission and distribution delays, as well as 1 ms access time for retrieving states from the shared memory; each server handles the incoming requests in its own buffer in FCFS manner and responds to the client on completing the corresponding request; the average response time and incoming request aggregate are recorded periodically to generate the test outcomes by comparing them to the dynamic thresholds fetched from established legitimate profiles.

The purpose of assuming both transmission and distribution delays to be 0 is to simply quantify the length of the whole detection phase (testing latency). With regard to the transmission delay, it can be large due to the geographical distances in large-scale distributed systems and possibly can bring up the testing latency if the client sends request in a stop-and-wait manner (it does not send a request until the previous requests are all responded, therefore the request rate is quite low and the length of each testing round is required to be longer), yet since the detection will be completed just in several rounds, such increases in the detection length are not significant. The assumption of 0 distribution delay is also practical, since the computational overheads for the testing matrix and dynamic thresholds are negligible by precomputing and fetching the results from the profiles. For the 1 ms state maintenance time, since all

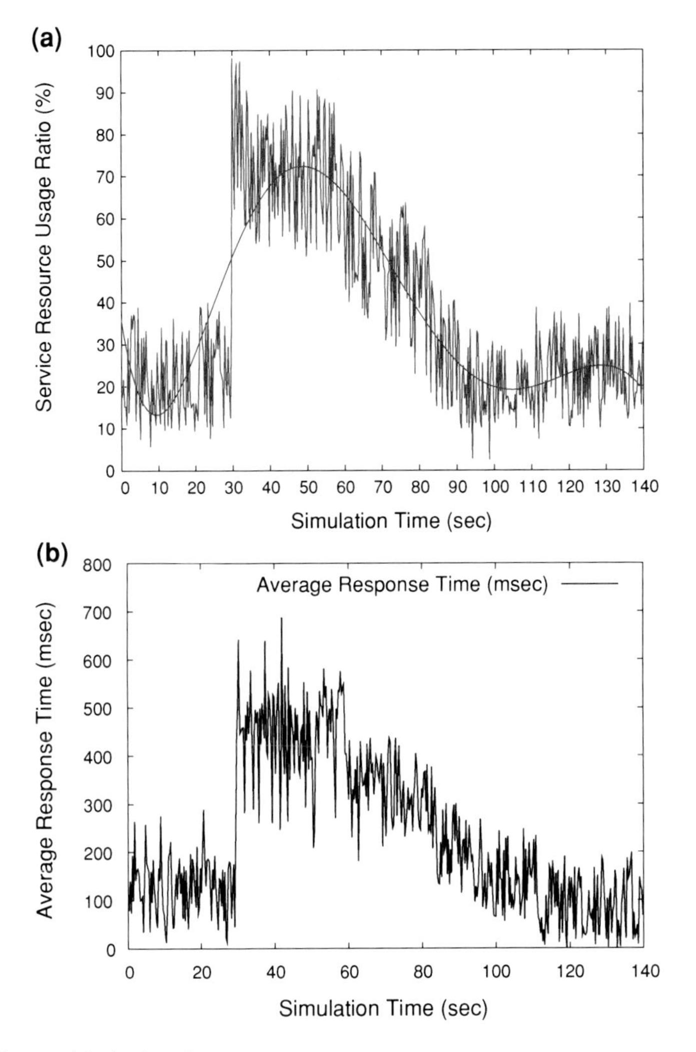

Fig. 2.6 Status of the back-end server

the clients are residing in one physical server, and all the virtual servers can quickly retrieve the client states from the shared memory, this is also a practical assumption.

With regard to the details of client behavior, the legitimate request interarrival rate is randomized from 1 to 3 request per ms, and the legitimate workload is randomized from 1 to 3 ms CPU processing time. On the contrary, the malicious request interarrival rates range from 5 to 20 per ms, and malicious workload range from 5 to 20 ms CPU time. Although requests with arbitrarily large rate or workload are favored by attackers, they are in fact easier to be discovered, so malicious requests with small margin from legitimate ones were also considered.

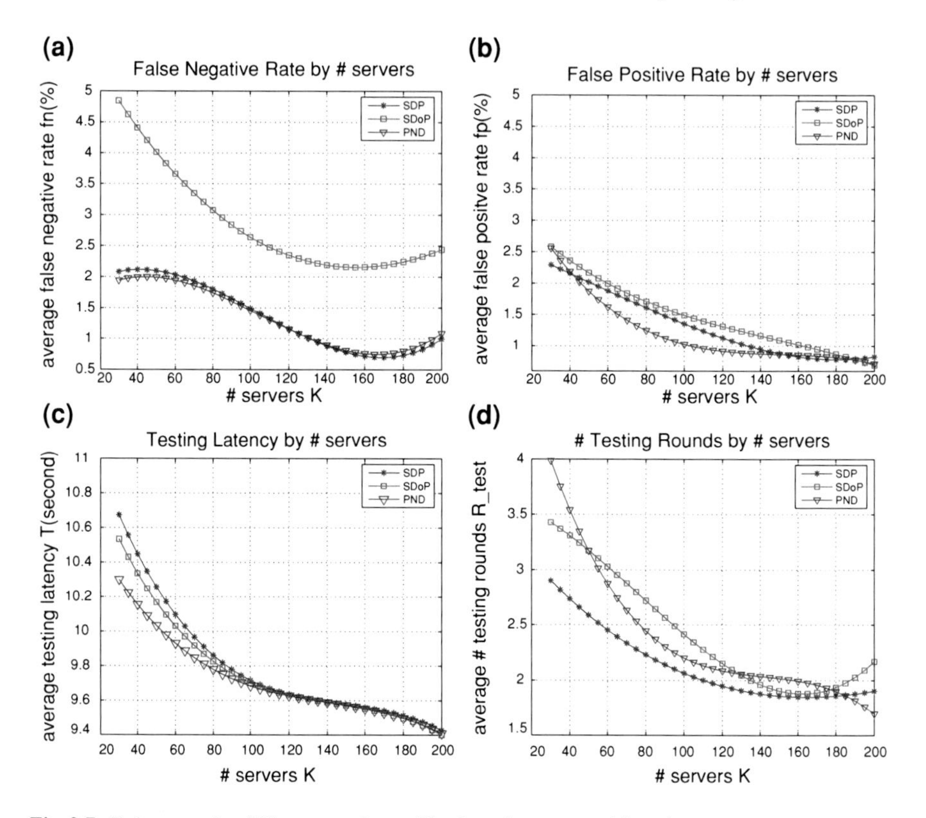

Fig. 2.7 Robustness by different numbers of back-end server machines k

2.6.2 Results

By setting $K = 50, n = 1,000, w = 100, d = 40, P = 1$ second, the efficiency of the GT-based detection system using *PND* algorithm, in terms of reducing the service resource usage ratio (*SRUR*) and the average response time (*ART*) are shown in Fig. 2.6a and b. The values of *SRUR* and *ART* climb up sharply at $t = 30$ s when the attack starts, and then gradually falls to normal before $t = 100$ s. Therefore, it takes only 70 s for the system to filter out attackers and recover to normal status. Note that actual detection latency should be shorter than this, because the threshold of *ART* for the system to convert from DANGER mode back to NORMAL mode is slightly higher than normal *ART*. Therefore, the system *SRUR* and *ART* will recover to normal shortly after the detection period ends.

In the following we show the robustness of the performance toward different environment settings: an increasing number of (1) virtual servers K; (2) malicious clients d; (3) all clients n.

In Fig. 2.7, a simulation of identifying $d = 10$ attackers out of $n = 1,000$ clients with the number of virtual servers ranging in $[17, 200]$ was studied. As can be seen, on one

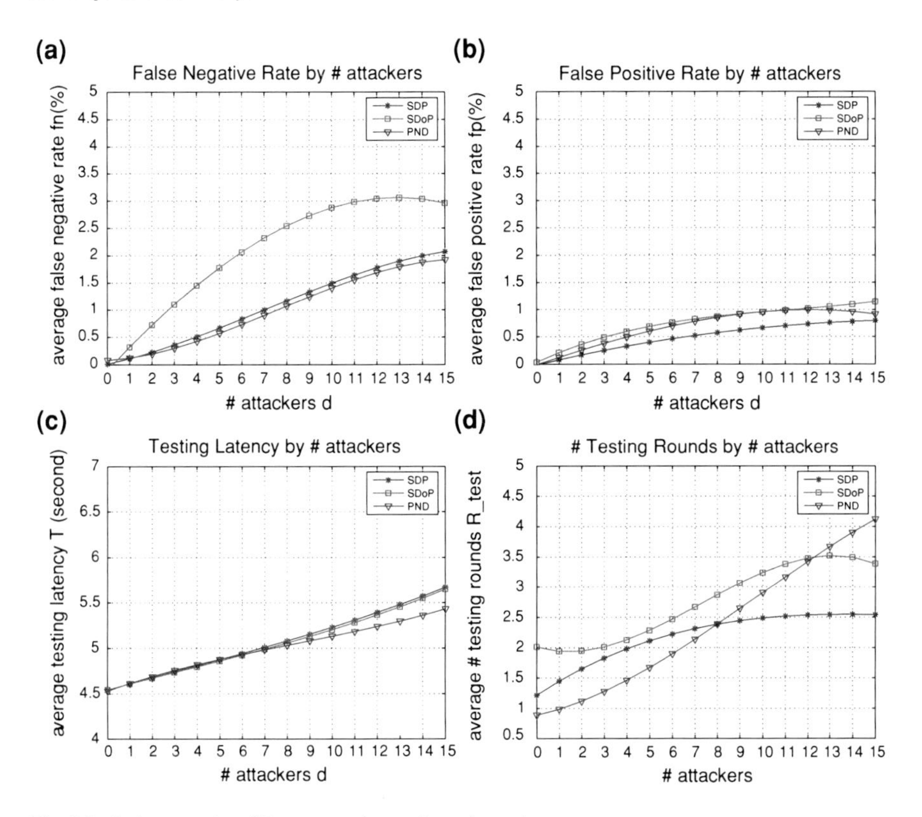

Fig. 2.8 Robustness by different numbers of attackers d

hand, all the false negative (Fig. 2.7a) and positive (Fig. 2.7b) rates are upper bound by 5% and decreasing as K goes up for all the three algorithms. This makes sense since the most possible case for an attacker to succeed in hiding itself is that it is accompanied by many clients with low request rate and workloads in a testing pool. In this case, it is quite possible that the aggregate interarrival rates and workloads in this server are still less than that of a server with the same number of legitimate clients. Therefore, the less clients serviced by each server, the less possibly this false negative case happens. Note that even if there is only one attacker but no legitimate client in a server whose incoming traffic is not quite dense, the server can still be tested as positive since all the tests are conducted based on some dynamic threshold (use the threshold for server with one client in this case), which changes as the number of active clients varies in the current virtual server. On the other hand, the testing latencies and the number of testing rounds keep declining from less than 11 s and 4 rounds respectively, which is because the identification will speed up with more available virtual servers.

With respect to the three different algorithms, they obtained approximate performances except that *SDoP* has slightly higher false negative rate than the other two.

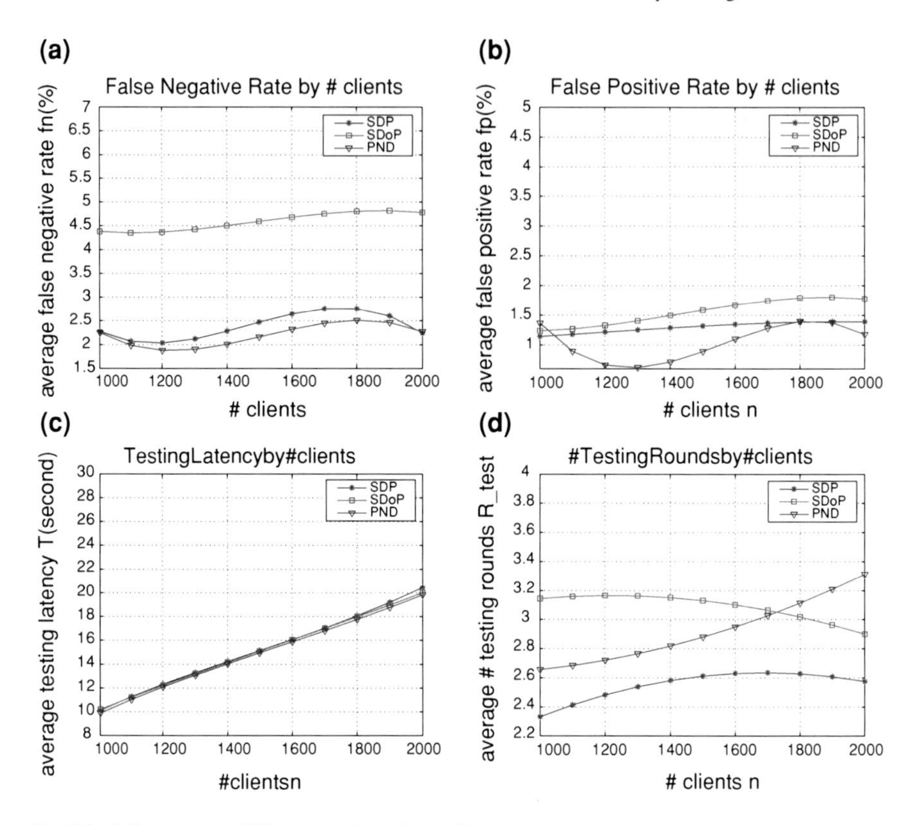

Fig. 2.9 Robustness by different total numbers of clients n

This is because of the legitimate clients identified at the earlier stages and staying in the servers till the end of the detection. Since their request rates and workloads are likely to be smaller than normal (that is why they are identified earlier), they may camouflage attackers in the following rounds.

In Fig. 2.8, the identification for $n = 1,000$, $K = 50$, $w = 100$, $P = 1$ s with [3, 17] attackers was conducted. As illustrated in Fig. 2.8, all the values of the four measures are slowly increased as d goes up. Overall, the false negative/positive rates are limited to less than 5% and the testing latencies are smaller than 6 s with 5 testing rounds. Similar to the previous experiments, *PND* and *SDP* exhibit better performances than *SDoP* in terms of false negative rate, but for the other measures, they are approximately the same.

Figure 2.9 shows its robustness for the cases $d = 10$, $K = 50$, $w = 100$, $P = 1$ s with [1,000, 2,000] clients. Apparently, both the false rates and the number of testing rounds keep stably below 5 % and 4 rounds respectively, toward increasing client amount. The testing latencies grow up from 10 to 20 s for all three algorithms, due to the increasing time costs for state maintenance toward a double number of clients (from 1,000 to 2,000). However, this small latency is still tolerable in real-time applications and can be further reduced by decreasing the state maintenance time cost within the same physical machine.

Overall, the simulation results can be concluded as follows:

- in general, the system can efficiently detect the attacks, filter out the malicious clients, and recover to NORMAL mode within a short period of time, in real-time network scenarios;
- all the three detection algorithms can complete the detection with short latency (less than 30 s) and low false negative/positive rate (both less than 5%) for up to 2,000 clients. Thus they are applicable to large-scale time/error-sensitive services;
- the *PND* and *SDP* algorithms achieve slightly better performance than the *SDoP* algorithm. Furthermore, the efficiency of the *PND* algorithm can be further enhanced by optimizing the *d*-disjunct matrix employed and state maintenance efficiency.

References

1. Sekar V, Duffield N, van der Merwe K, Spatscheck O, Zhang. H (2006) LADS: large-scale automated DDoS detection system. In: USENIX annual technical conference 2006
2. Kandula S, Katabi D, Jacob M, Berger AW (2005) Botz-4-sale: surviving organized DDoS attacks that mimic flash crowds 2nd NSDI. MA, Boston, May 2005
3. Ranjan S, Swaminathan R, Uysal M, Knightly E (2006) DDos-resilient scheduling to counter application layer attacks under imperfect detection. In Proceedings of the IEEE infocom, barcelona, Spain, April, 2006
4. Kim Y, Lau WC, Chuah MC, Chao HJ (2004) Packetscore: statisticsbased overload control against distributed denial-of-service attacks. In: Proceedings of infocom, HongKong, 2004
5. Du DZ, Hwang FK (2006) Pooling designs: group testing in molecular biology. World Scientific, Singapore
6. Atallah MJ, Goodrich MT, Tamassia R (2005) Indexing information for data forensics, ACNS. Lecture notes in computer science vol 3531. Springer, Heidelberg, pp 206–221
7. Ricciulli L, Lincoln P, Kakkar P (1999) TCP SYN flooding defense. In: Proceedings of CNDS
8. Gligor VD (2003) Guaranteeing access in spite of distributed service-flooding attacks. In: Proceedings of the security protocols workshop
9. Kargl F, Maier J, Weber M (2001) Protecting web servers from distributed denial of service attacks. In WWW '01: Proceedings of the 10th international conference on World Wide Web. ACM Press, New York, USA, pp 514–524
10. Thai MT, Xuan Y, Shin I, Znati T (2008) On detection of malicious users using group testing techniques. In: Proceedings of IEEE international conference on distributed computing systems (ICDCS)
11. Sharma P, Shah P, Bhattacharya S (2003) Mirror hopping approach for selective denial of service prevention in WORDS'03
12. Service provider infrastructure security: detecting, tracing, and mitigating network-wide anomalies (2005). http://www.arbornetworks.com 2005
13. Chu Y, Ke J (2007) Mean response time for a G/G/1 queueing system: simulated computation. Appl Math Comput 186(1):772–779
14. Eppstein D, Goodrich MT, Hirschberg D (2005) Improved combinatorial group testing algorithms for real-world problem sizes WADS. LNCS vol 3608. Springer, Heidelberg, pp 86–98

15. Mori G, Malik J (2003) Recognizing objects in adversarial clutter: breaking a visual captcha. IEEE Computer Vision and Pattern Recognition
16. Dyachkov AD, Rykov VV, Rachad AM (1989) Superimposed distance codes. Prob Control Inform Thy 18:237–250
17. Dyachkov AG, Macula AJ, Torney DC, Vilenkin PA (2001) Two models of nonadaptive group testing for designing screening experiments. In: Proceeding 6th International workshop on model-oriented designs and analysis. p 635

Chapter 3
Interference Free Group Testing and Reactive Jamming Attacks

Abstract Another application of group testing that we would like to introduce is to against reactive jamming attacks, which has emerged as a great security threat to wireless sensor networks, due to its mass destruction to legitimate sensor communications and difficulty to be disclosed and defended. There exist many studies against these attacks, however, these methods, i.e., frequency hopping or channel surfing, require excessive computational capabilities on wireless devices. To overcome the shortcomings, we present here an interference free group-testing based solution by identifying the trigger nodes, whose transmissions activate any reactive jammers. The identification of these trigger nodes can help us to (1) carefully design a better routing protocol by switching these nodes into only receivers to avoid activating jammers and (2) locate the jammers based on the trigger nodes, thus providing an alternative mechanism against reactive jamming attacks. The theoretical analysis and experimental results show that this advanced scheme performs well in terms of time and message complexities.

3.1 Overview

Since the last decade, the security of wireless sensor networks (WSNs) has attracted numerous attentions, due to its wide applications in various monitoring systems and vulnerability toward sophisticated wireless attacks. Among these attacks, jamming attack, where a jammer node disrupts the message delivery of its neighboring sensor nodes with interference signals, has become a critical threat to WSNs. As summarized in [1], various efficient defense strategies have been proposed and developed. However, a reactive variant of this attack, where jammer nodes stay quiet until an ongoing legitimate transmission (even has a single bit) is sensed over the channel, emerged recently and called for stronger defending systems and more efficient detection schemes.

M. T. Thai, *Group Testing Theory in Network Security*, SpringerBriefs in Optimization,
DOI: 10.1007/978-1-4614-0128-5_3, © My T. Thai 2012

Existing countermeasures against Reactive Jamming attacks consist of jamming (signal) detection and jamming mitigation.

On the one hand, detection of interference signals from jammer nodes is non-trivial due to the discrimination between normal noises and adversarial signals over unstable wireless channels. Numerous attempts to this end monitored critical communication-related objects, such as *Receiver Signal Strength* (RSS), *Carrier Sensing Time* (CST), and *Packet Delivery Ratio* (PDR), compared the results with specific thresholds, which were established from basic statistical methods and multi-modal strategies [1, 2]. By such schemes, jamming signals could be discovered; however, how to locate and catch the jammer nodes as well as to avoid the jamming based on these signals are much more complicated and have not been settled.

On the other hand, various network diversities are investigated to provide mitigation solutions [3–6]. Spreading spectrum [1, 7, 8] exploiting the properties of multiple frequency bands and MAC channels, channel surfing [9], multi-path routing [3] are some good examples. In these methods, the capability of jammers are assumed to be limited and powerless to catch the legitimate traffic from the camouflage of these diversities. Unfortunately, due to the silent behavior of reactive jammers, they have more powers to destruct these mitigation methods. In addition, due to lack of pre-knowledge over possible positions of hidden reactive jammer nodes, legitimate nodes cannot efficiently evade jamming signals, especially in the dense sensor networks when multiple mobile nodes can easily activate reactive jammer nodes and cause the interference. To this end, other solutions are in great need.

Recently, a mapping service of jammed area has been introduced [10], which detects the jammed areas and suggests that routing paths evade these areas. This method is suitable for proactive jamming, since all the jammed nodes are having low PDR and thus incapable for reliable message delay. However, in the case of reactive jamming, as will be shown later, this is not always the case. This approach can create unnecessarily big jammed region and result in isolated networks in the worst case. Additionally, the message overhead is relatively high during the mapping processing.

To overcome the mentioned shortcomings, a novel method against reactive jamming attack in multi-channels WSNs based on the group testing technique was introduced [11, 12]. In this method, the trigger nodes whose broadcasting activate the reactive jammers are identified by using GT. The identification of *trigger* nodes can have several benefits against reactive jamming attacks. First of all, we can construct a routing algorithm in which the triggers are only receivers, thus avoiding activating the jammers and minimizing the effect of jamming attacks. In case where a trigger node needs to send a message, we may still utilize the use of channel surfing, however, only a few nodes may require this operation, thus greatly reducing the computational costs required by existing methods. In addition, the identification of trigger nodes will not create an unnecessary big jammed region as in [10].

Furthermore, after the identification of trigger nodes, victim nodes (VN) would be scheduled to transmit messages in order to minimize the damage from the attackers by keeping silent during the transmission of the trigger nodes, thus preventing from being jammed.

Although the benefits of this trigger-identification approach are exciting, its hardness is also obvious, which dues to the efficiency requirements of identifying the set of trigger nodes out of a much larger set of VNs that are affected jamming signals from reactive jammers with possibly various sophisticated behaviors. Thus group testing comes to the picture for help.

In the context of trigger nodes identification, all the *victim* nodes which are within the transmission range of the jammers are first identified using some existing methods such as alarm message or investigating corresponding links' PDR and RSS, then these victim nodes are pooled into multiple interference free *testing teams* and all the victim nodes within each testing team further grouped into several *groups* based on a d-disjunct matrix to transmit out testing packets. A group with no noise heard is called negative group, which means all the VNs within this group cannot trigger any jammers. On the contrary, a group of hearing noises are called positive group, which means at least one of the VNs is a trigger node, thus requires further tests on this group until the trigger nodes are identified. Therefore, the realization of this solution based on GT must address the following issues: (1) how to properly pool the VNs into multiple interference free testing teams so that different teams cannot interfere each others test outcome (i.e., a VN triggers the activation of a jammer, whose noise reaches another team's hearing nodes, in which case, the latter team will have positive outcome even if it has no trigger node); (2) estimate the upper bound d of the number of trigger nodes in order to construct a corresponding d-disjunct matrix and (3) how to collect and decode the outcomes.

By utilizing the GT theory, disk-cover-based grouping, and clique-based clustering, this advanced solution can accurately identify the trigger nodes among the VNs with low message and computational complexity. This is critical and suitable for WSNs since they have limited resources and energy conservation. Detailed theoretical analysis and simulation results show the novel performance of this protocol in terms of time and message complexity.

Before introducing the solutions to the above three problems, we first present the preliminaries in the following section.

3.2 Problem Models and Preliminaries

3.2.1 Network Model

We consider a wireless sensor network consisting of n sensor nodes and one base station. Each sensor node is equipped with a globally synchronized time clock, omnidirectional antennas, and m radios for in total k channels throughout the network, where $k > m$. For simplicity, the power strength in each direction is assumed to be uniform, so the transmission range of each sensor can be abstracted as a constant r_s and the whole network as a *unit disk graph* (UDG) $G = (V, E)$, where any node pair i, j is connected iff the Euclidean distance between i, j: $\delta(i, j) \leq \mathrm{r}_s$.

3.2.2 Basic Attacker Model

Conventional reactive jammers [1] are defined as malicious devices, which keep idle until they sense any ongoing legitimate transmissions and then emit jamming signals (packet or bit) to disrupt the sensed signal (called jammer wake-up period). Once the sensor transmission finishes, the jamming attacks will be stopped (called jammer sleep period). Three concepts are introduced to complete this model.

Jamming range R. Similar to the sensors, the jammers are equipped with omni-directional antennas with uniform power strength on each direction. The jammed area can be regarded as a circle centered at the jammer node, with a radius R, where R is assumed greater than r_s, for simulating a powerful and efficient jammer node. All the sensors within this range will be jammed during the jammer wake-up period. The value of R can be approximated based on the positions of the *boundary sensors* (whose neighbors are jammed but themselves not), and then further refined.

Triggering range r. On sensing an ongoing transmission, the decision whether or not to launch a jamming signal depends on the power of the sensor signal P_s, the arrived signal power at the jammer P_a with distance r from the sensor, and the power of the background noise P_n.

According to the traditional signal propagation model, the jammer will regard the arrived signal as a sensor transmission as long as the Signal-Noise Ratio is higher than some threshold, i.e., $SNR = \frac{P_a}{P_n} > \theta$ where $P_\mathrm{a} = \frac{P_s}{r^\xi} \cdot Y$ with θ and ξ called jamming decision *threshold* and *path-loss* factor, Y as a log-normally random variable. Therefore, $r \geq (\frac{\theta \cdot P_\mathrm{n}}{P_s} \cdot Y)^{\frac{1}{\xi}}$ is a range within which the sensor transmission will definitely trigger the jamming attack, named as *triggering range*. As will be shown later, this range r is bounded by R from above, and r_s from below, where the distances from either bounds are decided by the jamming decision threshold θ. For simplicity, we assume triggering range is the same for each sensor. In this chapter, we also consider the triggering range is equal to the transmission range, that is, $r = \mathrm{r}_s$. We will investigate the case $r \neq \mathrm{r}_s$ in Chap. 4.

Jammer distance. Any two jammer nodes are assumed not to be too close to each other, i.e., the distance between jammer J_1 and J_2 is $\delta(J_1, J_2) > R$. The motivations behind this assumptions are threefold: (1) the deployment of jammers should maximize the jammed areas with a limited number of jammers, therefore large overlapping between jammed areas of different jammers lowers down the attack efficiency; (2) $\delta(J_1, J_2)$ should be greater than R, since the transmission signals from one jammer should not interfere the signal reception at the other jammer. Otherwise, the latter jammer will not be able to correctly detect any sensor transmission signals, since they are accompanied with high RF noises, unless the jammer spends a lot of efforts in de-noising or embeds jammer-label in the jamming noise for the other jammers to recognize. Both ways are infeasible for an efficient attack; (3) the communications between jammers are impractical, which will expose the jammers to anomaly detections at the network authority.

3.2.3 Maximum Clique

The Maximum Clique problem is defined as follows. Given an undirected graph $G = (V, E)$, the MC problem asks us to find a subgraph G with maximum size such that G' is a clique. A graph is said clique if all of its vertices are pairwise adjacent. The maximum clique problem is one of the first problems shown to be **NP**-complete [13].

So far, the best polynomial-time approximation algorithm for the maximum clique problem was developed by Boppana and Halldorsson and achieved an approximation ratio of $n^{(1-o(1))}$ [13]. Its inapproximability was later shown to be $n^{1-\varepsilon}$ by Hastad [13]. There are some other results in the literature concerning the approximation of the maximum clique problem on arbitrary or special graphs [13–15].

In this chapter, the maximum clique problem is applied to obtain the upper bound of the number of trigger nodes based on the number of reactive jammers. Since a jammer can only be activated by the nodes within a certain distance, we can construct a unit disk graph of all nodes with the radius twice the distance to estimate the upper bound of the number of trigger nodes.

3.3 Group-Testing-Based Trigger Node Identification: Preprocessing

3.3.1 Node Classification

Before starting the testing procedure to identify the trigger nodes, we first classify the nodes into three classification: (1) Victim Nodes (VN): if a node v hears a jamming signals, then v is VN; (2) Unaffected Nodes (UN): Nodes are not being jammed, i.e., cannot hear the jamming signals; (3) Boundary Nodes (BN): v is said a BN if v is an UN who has a neighbor node as a victim. The purpose of BNs is to estimate the jamming range R where the VN set is to be tested to obtain the trigger nodes. The relationships among these classes are shown in Fig. 3.1.

Each node can locally classify itself into the corresponding classification based on the following simple procedure. Each node periodically sends a status report message to the base station. When generating the status report message, each sensor can locally obtain its jamming-status and either label itself as an UN, a VN, or a BN. In detail, if a node v hears jamming signals, it will not try to send out messages but keep its label as *victim*. If v cannot sense jamming signals, its report will be routed to the base station as usual; however, if it does not receive ACK from its neighbor on the next hop of the route within a timeout period, it tries for two more retransmissions. If no ACKs are received, it is quite possible that neighbor is a VN, then v labels itself as boundary. All the other nodes are regarded as UNs.

The base station waits for the status report from each node in each period of length $\mathscr{P}$. If no reports have been received from a node v with a *maximum delay time*, then

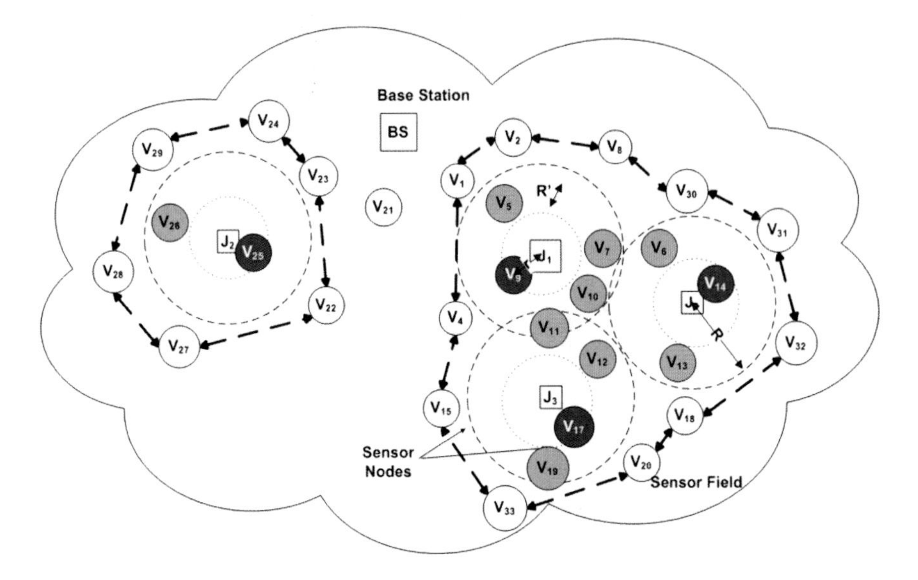

Fig. 3.1 Nodes in *grey* and *blue* are VNs around jammer nodes, where *blue nodes* are also trigger nodes, which invoke the jammer nodes. Nodes surrounding the jammed area are BNs, while the others are UNs

v will be regarded as victim. The maximum delay time is related with graph diameter and will be specified later. If the aggregate report amount is less than ψ, the base station starts to create the testing schedule for the trigger nodes, based on which the routing tables will be updated locally. Details of this implementation can be founded in [16].

3.3.2 Jamming Range Estimation

The jamming range R can be estimated based on the locations of the boundary and VNs.

In the *sparse-jammer* case where the distribution of jammers is relatively sparse and there is at least one jammer whose jammed area does not overlap with the others, like J_2 in Fig. 3.1. By denoting the set of BNs for the ith jammed area as BN_i, the coordinate of this jammer can be estimated as

$$(X_J, Y_J) = (\frac{\sum_{k=1}^{BN_i} X_k}{|BN_i|}, \frac{\sum_{k=1}^{BN_i} Y_k}{|BN_k|})$$

where (X_k, Y_k) is the coordinate of a node k is the jammed area BN_i and then further the jamming range R can be estimated as

$$R = \min_{\forall BN_i}\{\max_{k \in BN_i}(\sqrt{(X_k - X_J)^2 + (Y_k - X_J)^2})\}$$

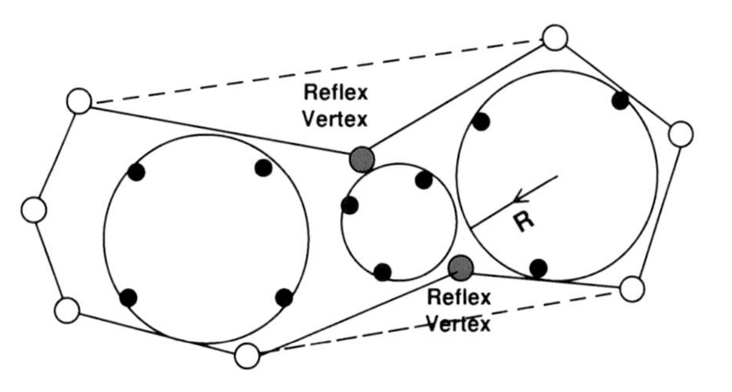

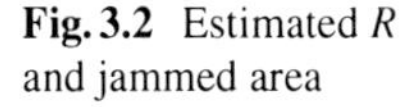
Fig. 3.2 Estimated R and jammed area

since all the jammers have the same range.

Otherwise in the *dense-jammer* case, as shown in Fig. 3.2, we need to first estimate the jammed areas, which are simple polygons (unnecessarily convex) containing all the boundary and VNs. This process consists of three steps: (1) Discovery of *convex hulls* of the boundary and VNs, where no UNs are included in the generate convex polygons. (2) For each boundary node v not on the hull, choose two nodes on the hull and connect v to them in such a way that the internal angle at this reflex vertex is the smallest, hence the polygon is modified by replacing an edge (dotted one in Fig. 3.2) by the two new ones. The resulted polygon is the estimated jammed area. (3) Execute the near-linear algorithm [17] to find the optimal variable-radii disk cover of all the VNs, but constrained in the polygon, and return the largest disk radius as R.

3.4 Identifying Trigger Nodes Algorithm

We are now ready to present the main algorithm, called Identifying Trigger Nodes (ITN), of the group-testing-based trigger-identification method. As briefly mentioned earlier, the basic idea that ITN algorithm uses is as follows: we test victims to identify the trigger nodes by two steps: (1) We use Interference Free Group Testing (IFGT) to divide victim nodes into multiple interference free testing teams based on the minimum collection of disjoint disk cover . Each cover includes a set of disjoint disk where the center of this disk will act as a outcome collector. Each of the disjoint disks can be tested simultaneously. For a set of victims in each testing team (i.e., in each disjoint disk), we use the Non-adaptive Group Testing Detection (NGTD) algorithm to detect all trigger nodes within these VNs. We continue covering and testing VNs until all VNs are tested.

3.4.1 *Interference Free Group Testing Algorithm*

The goal of this process is to partition VNs so as to simultaneously test as many interference free testing team as possible. Recall that two testing teams are interference free iff the testing result of this team cannot interfere the testing result of the other team. That said, if a trigger in this testing team activates a jammer A, this jammer A cannot jam any sensor nodes in the other testing team.

The basic idea of this algorithm is as follows. For each VN v, construct two disks D_v^1 and D_v^2 centered at v with radius $(R-r)$ and $(3R-r)$, respectively. The objective is to obtain a minimum collection of disjoint disk covers, where each cover is a set of disjoint disks such that VNs within each disjoint disk can be tested simultaneously without mutual interference. We adopt a greedy method for this by selecting a node v whose corresponding $R-r$ disk covers the maximum number of VN, and then selects another node u similarly after removing all the nodes within $3R-r$ from v from the graph. Iterate this until no nodes are left in the graph. The details of IFGT are shown in Algorithm 7.

Algorithm 7 The *IFGT* Algorithm

1: **Input**: All left victim nodes W_{i-1} after cover $i-1$
2: **Output**: The collection of groups in all covers $G_{i1}, \ldots, G_{ij}$, $1 \leq i \leq \mathscr{C}$, $1 \leq j \leq \sqcup_i$
3: $i \leftarrow 1$
4: **while**$|W| \neq 0$ **do**
5: ▷ Construct double disks for each victim node
6: **for** $w \in W_{i-1}$ **do**
7: Construct D_w^1 and D_w^2
8: **end for**
9: $k \leftarrow 1$
10: $W_i \leftarrow \emptyset$
11: **while**$|W_{i-1}| \neq 0$ **do**
12: Choose $w \in W_{i-1}$ to maximize $\kappa(D_w^1)$
13: $G_{ik} \leftarrow D_w^1$
14: $W_{i-1} \leftarrow W_{i-1} \setminus D_w^2$
15: $W_i \leftarrow W_i \cup \{D_w^2 \setminus D_w^1\}$
16: $k \leftarrow k+1$
17: **end while**
18: $W \leftarrow W_i$
19: **end while**

3.4.2 Non-Adaptive Group Testing Detection Algorithm

After finding the collection of disjoint disk covers, we start conducting the group testing for each cover. Notice that for each cover which consists of a set of disjoint disks (i.e., a set of interference free testing teams), we will conduct the test for each disk simultaneously. For each disk j in cover i, all the victims in this disk are tested based on the $\boldsymbol{NGTD}$ algorithm to detect all trigger nodes. In order to apply the non-adaptive group testing technique, we need to estimate an upper bound D_{ij} of the number of trigger nodes in each disk. Based on this obtained value D_{ij}, NGTD algorithm constructs a D_{ij}-disjunct matrix accordingly using Algorithm 2. From this matrix, the algorithm will partition VNs into several groups and start the identification procedure as shown in Algorithm 8.

Algorithm 8 The NGTD Algorithm on group j in cover$i$

1: **Input**: Victim nodes set W_{ij} in one group, R, r
2: **Output**: Number of trigger nodes D_{ij} in this group
3: Construct $G_{ij} = (W_{ij}, E_{ij})$, where $E_{ij} = \{(u, v)|\delta(u, v) \leq 2r, u, v \in W_{ij}\}$
4:
5: ▷ Find the upper bound D_{ij}
6: $D_{ij} \leftarrow 0$
7: **for** $k = 1$ to $|J_{ij}|$ **do**
8: Find the MAXIMUM CLIQUE $c(G_{ij})$ on graph G_{ij}
9: $G_{ij} \leftarrow G_{ij} \setminus \bigcup_{w \in c(G_{ij})} w$
10: $D_{ij} \leftarrow D_{ij} + |c(G_{ij})|$
11: **end for**
12: ▷ Test by using NON- ADAPTIVE GT
13: Construct a D_{ij}-DISJUNCT MATRIX M_{ij} using Algorithm 2
14: Group the column in each row with entity 1 into one group
15: Test these groups simultaneously
16: Decode the testing result to identify all trigger nodes

The complete *ITN* algorithm is presented in Algorithm 9.

3.5 Theoretical Analysis

3.5.1 Estimation of Trigger Node Upper Bound D_{ij}

In this section, we present the estimation on the upperbound D_{ij}, and the number of the trigger nodes, in each disk. For simplicity, we assume that the interference radius is larger than legitimate transmission radius, $R = \alpha r$ where $\alpha > 1$ since jammers have more capabilities than normal sensor nodes. Let J be the set of jammers that a trigger node t could activate. We note that the distances from jammers in J to t are at most r while the distance between any two jammers must be larger than $R = \alpha r$; otherwise jammers will invoke each other and run out of energy. We have the following lemma:

Algorithm 9 The *ITN* Algorithm

1: **Input**: A WSN $G = (V, E)$
2: **Output**: The set U of all trigger nodes
3: $W \leftarrow$ The set of victim nodes
4: $W_i \leftarrow$ The set of left victim nodes from cover $i - 1$
5: $G_{ij} \leftarrow$ The group j in cover i
6: $U_i \leftarrow$ The set of trigger detected in cover i
7: $W \leftarrow \emptyset$, $U \leftarrow \emptyset$, $W \leftarrow$ Victim nodes
8: $W_1 \leftarrow W$
9: $i = 1$
10: **while**$|W_i| > 0$ **do**
11: $\quad G_{ij} \leftarrow$ Partition based on the *IFGT* algorithm in cover i
12: $\quad U_i \leftarrow$ trigger nodes based on the *NGTD* algorithm
13: $\quad U \leftarrow U \cup U_i$
14: $\quad i \leftarrow i + 1$
15: **end while**

Lemma 3.1 *Let J be the set of jammers that a trigger node t could activate, then* $|J| < \frac{\pi}{\arcsin(\frac{\alpha}{2})}$

Proof Let O be the location of t. Assume that J contains jammers with locations $J_1, J_2, \ldots, J_m$ in clockwise order like in Fig. 3.3, where $m = |J|$. We have $OJ_i (= \delta(O, J_i)) \leq r \forall i = 1 \ldots m$ and $J_i J_j > R = \alpha r \forall 1 \leq i < j \leq m$.

Since $\sum_{i=1}^{m} \widehat{J_i O J_{i+1}} = 2\pi$ where $J_{m+1} \equiv J_1$, let $\widehat{J_i O J_{i+1}} = \beta$ be the smallest angle, we have $\beta \leq \frac{2\pi}{m}$ and $|J| < \frac{2\pi}{\beta}$.

Use the cosine's law: $R^2 < J_i J_{i+1}^2 = OJ_i^2 + OJ_{i+1}^2 - 2OJ_i OJ_{i+1} \cos\beta$

As jammers will not revoke each other, we have $J_i J_{i+1} > R > r > \max\{OJ_i, OJ_{i+1}\}$. Hence, $J_i O J_{i+1}$ will be the largest angle of the triangle $J_i O J_{i+1}$. We obtain $\beta > \frac{2\pi}{6}$ i.e., $m < 6$.

From $\beta > \frac{2\pi}{6}$, $OJ_i^2 + OJ_{i+1}^2 - 2OJ_i OJ_{i+1} \cos\beta$ obtains the maximum value at $OJ_i = OJ_{i+1} = r$. Hence, $\alpha^2 r^2 < r^2(2 - 2\cos\beta)$ or $\beta > \arccos(1 - \frac{\alpha^2}{2}) = 2\arcsin(\frac{\alpha}{2})$. Therefore, $|J| = m < \frac{2\pi}{2\arcsin(\frac{\alpha}{2})}$. □

Following Lemma 3.1, we have:

- $|J| \leq 1$ when $\alpha \geq 2$.
- $|J| \leq 2$ when $\alpha \geq \sqrt{3}$.
- $|J| \leq 3$ when $\alpha \geq \sqrt{2}$.
- $|J| \leq 4$ when $\alpha \geq \sqrt{\frac{5-\sqrt{5}}{2}}$.
- $|J| \leq 5$ when $\alpha \geq 1$.

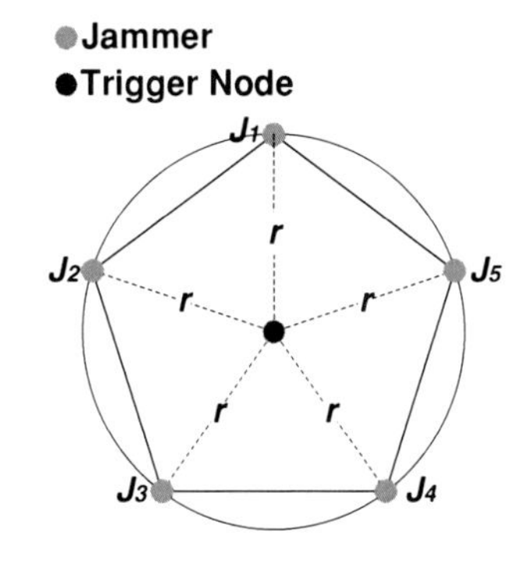

Fig. 3.3 Five possible jammers activated by a trigger node t

Theorem 3.1 *The upper bound D_{ij} of the number of trigger nodes in one disk is*

$$\left|\bigcup_{k=1}^{J_{ij}} c_k(G_{ij})\right|$$

where $c_k(G)$ is the kth maximum clique on graph G.

Proof According to Lemma 3.1, the nodes in disk j of cover i can trigger at most $|J_{ij}|$ jammers. Intuitively, we know that a set of trigger nodes to activate the same jammer have less than distance $2r$. In Algorithm 8, we construct a unit disk graph $G_{ij} = (W_{ij}, E_{ij})$ with disk radius $2r$ so that the nodes which trigger the same jammer must form a clique in graph G_{ij}.

In each iteration s, according to Algorithm 8, we choose the first J_{ij}^{th} maximum cliques and unite all these cliques. That is, $\left|\bigcup_{k=1}^{J_{ij}} c_k(G_{ij})\right|$, completing the proof. □

3.5.2 Correctness of ITN Algorithm

Lemma 3.2 *Any two nodes with the distance larger than $R+r$ are interference-free nodes.*

Proof Assume that any two nodes u and v with distance $\delta(u, v) > R + r$ are not interference free. Then there exists a jammer J such that J can interfere both u and v. Without lost of generality, we can assume that node v activates J. Thus $\delta(v, J) \leq r$. Plus,$\delta(v, J) \leq R$ and $\delta(u, J) \leq R$, then $\delta(u, v) \leq R + r$, which contradicts to our assumption. □

Lemma 3.3 *For each set of VN in disk D_v^1, the center node v can be used as the collector. That is v can sense the noises from any jammers triggered by any nodes within the distance $R - r$ from v.*

Proof The proof is straightforward. Assume that a center node v in disk D_v^1 cannot sense the noise from a jammer J, which is activated by a node u in the disk D_v^1. Then, we have $\delta(u, J) \leq r$ and $\delta(v, J) > R$. Therefore, $\delta(u, v) > R - r$, contradicting to the fact that u is in D_v^1. □

Theorem 3.2 *The ITN algorithm can correctly identify all trigger nodes.*

Proof Since the jammer noise range R is always larger than normal transmission range r, the trigger nodes must be included in the VN. Therefore, if we test all victim nodes, we must be able to identify all trigger nodes. Note that from Lemma 3.2 and the fact that each disk D_v^1 has a radius $(R-r)$, all the VN in any two different disjoint disks are interference-free. Thus the testing result is correctly collected. □

3.5.3 Performance Analysis

Lemma 3.4 *Given the UDG H with radius $3R-r$, denote the maximum node degree in H as $\Delta(H)$, the total number of rounds needed to cover all the sensor nodes is at most $\Delta(H)$.*

Proof This bound can be obtained by considering the disks in the same round as an MIS (Maximal Independent Set). For the ith round, denote the maximum node degree of the current graph H as Δ_i, and the set of any center u of the selected disks D_u^2 form an MIS of H, then the size of such an MIS is lower bounded by $\frac{|W_i|}{\Delta_i+1}$, where $|W_i|$ refers to the number of uncovered nodes at the beginning of this round. Henceforth, the number of nodes covered in the ith round at least equals to the size of this MIS, i.e., $\frac{|W_i|}{\Delta_i+1}$. Since the number of uncovered nodes is decreasing round by round, Δ_i is non-increasing for each round, so straightforwardly at most $\Delta(H)+1$ rounds where $\Delta(H) = \Delta_1 = \max_i \Delta_i$. □

Lemma 3.5 *The number of testing covers to detect trigger nodes in each group of VNs n_{ij} is upper bounded by*

$$\lceil \min\{(1+o(1))\frac{D_{ij}^2 \log_2^2 n_{ij}}{\log_2^2(D_{ij}\log_2 n_{ij})}, n_{ij}\}/m\rceil$$

where $D_{ij} = \left|\bigcup_{k=1}^{J_{ij}} c_k(G_{ij})\right|$.

Proof In WSNs, as we defined there are m radios so that at most m groups can be tested at the same time. The proof follows immediately from Theorem 1.5. □

Corollary 3.1 *The total number of testing rounds in cover C_i is upper bounded by*

$$\max_j \lceil \min\{(1+o(1))\frac{D_{ij}^2 \log_2^2 n_{ij}}{\log_2^2(D_{ij}\log_2 n_{ij})}, n_{ij}\}/m\rceil$$

Theorem 3.3 *The total testing round* $\mathcal{T}$ *is upper bounded by*

$$\sum_{i=1}^{\Delta(H)+1} \max_j \lceil \min\{(1+o(1))\frac{D_{ij}^2 \log_2^2 n_{ij}}{\log_2^2(D_{ij}\log_2 n_{ij})}, n_{ij}\}/m \rceil$$

where $D_{ij} = \left|\bigcup_{k=1}^{J_{ij}} c_k(G_{ij})\right|$.

Proof According to Lemma 3.5 and Corollary 3.1, the covers for all VN are $\Delta(H)+1$ and the testing time for each cover is the maximum testing time among all groups, that is,

$$\max_j \lceil \min\{(1+o(1))\frac{D_{ij}^2 \log_2^2 n_{ij}}{\log_2^2(D_{ij}\log_2 n_{ij})}, n_{ij}\}/m \rceil$$

where $D_{ij} = \left|\bigcup_{k=1}^{J_{ij}} c_k(G_{ij})\right|$. The proof is complete. □

3.6 Experimental Analysis

In this section, the efficiency of ITN algorithm through a series of simulations is evaluated. The results of these experiments show that ITN is timely efficient for ITN and defending reactive jamming attacks.

3.6.1 Simulation Setup

In order to simulate a general sensor network, N sensor nodes are randomly distributed along with one base station and J jammers to a square network field with width s. As mentioned above, the base station, sensor nodes, and jammers have transmission range, ρ, r, and R respectively. In order not to exaggerate the power of the base station, ρ is set to equal to r in this simulation, while larger ρ would make this solution more efficient.

There are total six benchmarks used in the simulations with different input parameter teams. On one hand, we study the average *number of disk covers* T in the *IFGT algorithm*, and the *maximum node degree* Δ to validate the bound of T proved in Theorem 3.1. On the other hand, we confirm the *overall test length* (number of rounds t) as analyzed in Theorem 3.3. Moreover, we record the *number of VNs* n and the *total volume of communication messages* M between the sensors and the base station, to indicate the message complexity of this method.

To investigate the effects of a series of network parameters over the efficiency of this solution, the values for the *number of jammers* J, *number of radios* m, *number of sensor nodes* N, *width of the square network region* s as well as the *noise range*

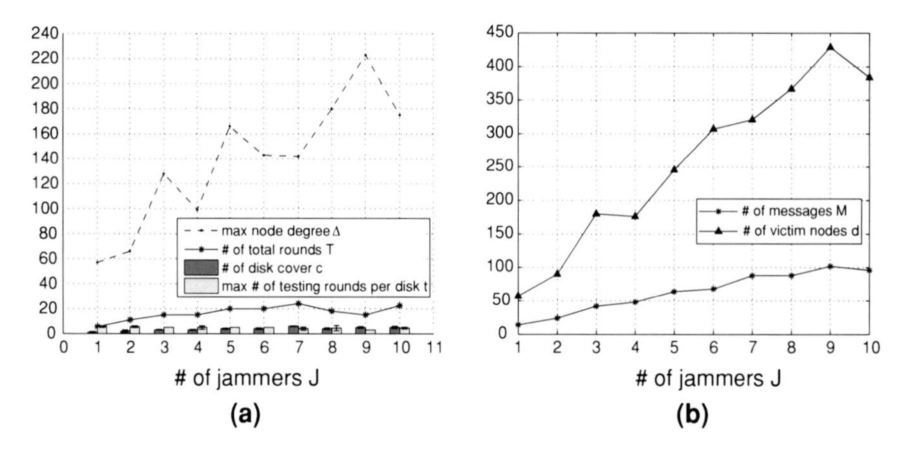

Fig. 3.4 Experimental results by varying the number of jammers

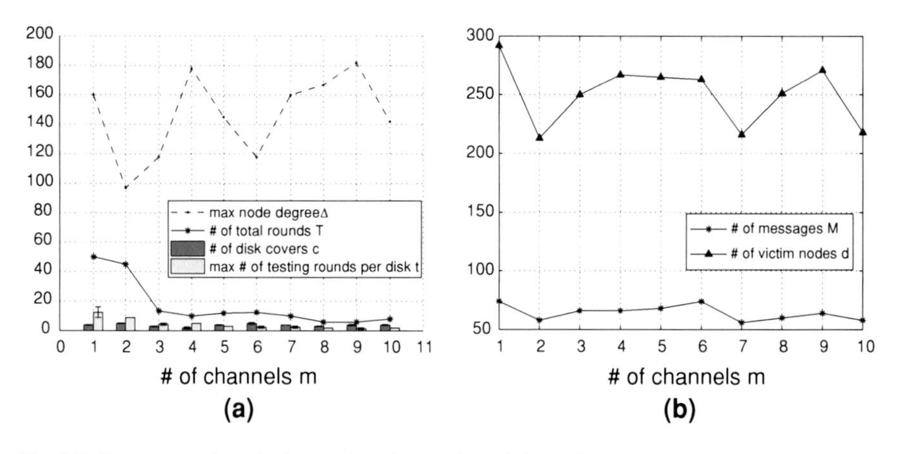

Fig. 3.5 Experimental results by varying the number of channels

ratio α were varied, hence the following subsections and Figs. 3.4, 3.5, 3.6, 3.7, and 3.8 are the corresponding results and analysis. Note that for each parameter team, 100 network instances are investigated and the results were averaged.

3.6.2 Results and Analysis

Performance by the number of jammers J. In this evaluation, $N = 1000$ nodes with $m = 3$ radios where $J \in [1, 10]$ jammers are randomly deployed on a 1500×1500 network. As can be seen in Fig. 3.4, the number of testing rounds T is very stable during the increase of J and N. The number of disk covers c and maximum number of testing rounds per disk t are smaller than 10, where the latter is much smaller

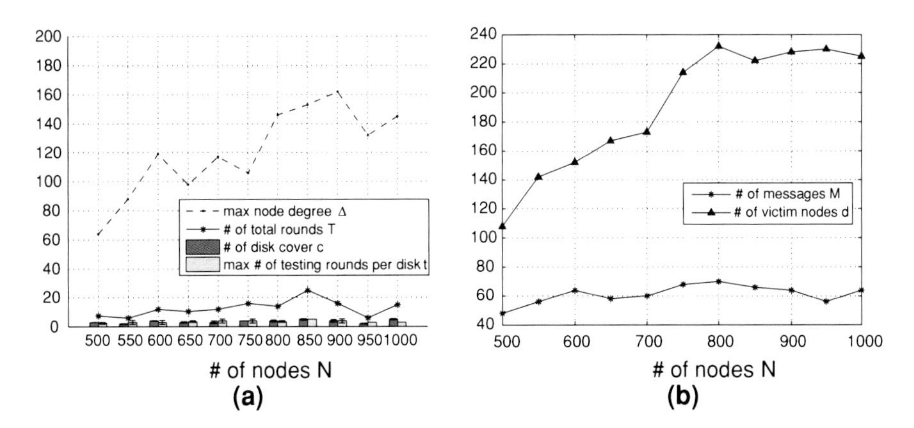

Fig. 3.6 Experimental results by varying the number of nodes in the network

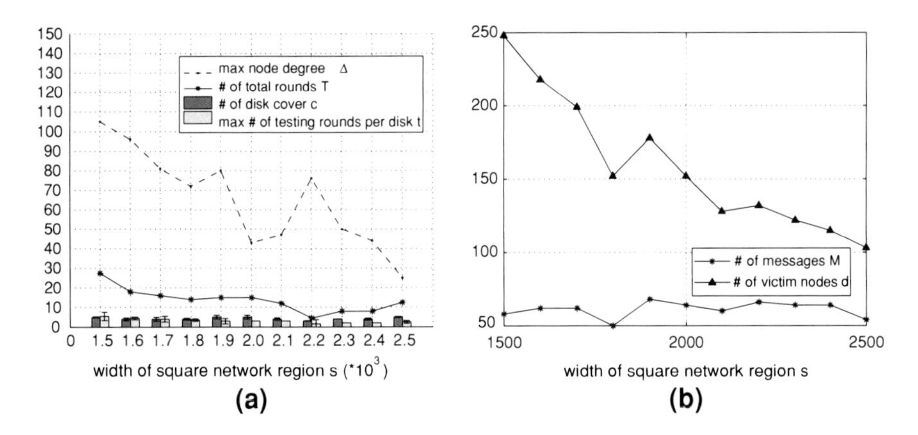

Fig. 3.7 Experimental results by decreasing network densities

than the maximum node degree Δ. This contributes to dramatically small number of overall rounds T, which is no larger than 30 and stable for increasing J.

Moreover, since each $(R - r)$-disk in our tests needs only one sensor node to send the result back to the base station, the message complexity M is also much smaller (less than 100) than the number of VNs n. Note that if individual testing was performed, M should be as high as $O(n)$. Therefore, ITN can promptly defend a jamming attack with increasing number of jammers, in terms of time complexity and message complexity.

Performance by the number of radios m. In this study, $N = 1000$ nodes and $J = 5$ jammers where $m \in [1, 10]$ were randomly distributed on a 1500×1500 network area. As in each testing team, we construct m pools and test them in parallel, therefore, the number of tests within an $(R - r)$-disk can be reduced by the factor m, which shortens the *overall test length*. As illustrated in Fig. 3.5a, the maximum testing rounds per disk decreases as the radio size increases, which assists to drop

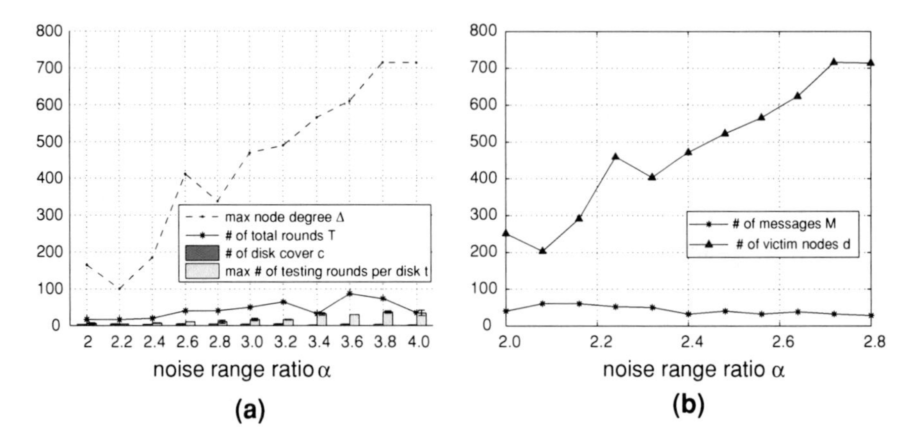

Fig. 3.8 Experimental results by varying the value of α

the total rounds T drastically. Especially, when m changes from two to one as shown in the Fig. 3.5a, the overall total rounds drop rapidly. Thus clearly the radio size can highly benefit the overall test length of the detection method.

Performance by the number of nodes N. In this study, we investigate the scalability of the proposed solution. As shown in Fig. 3.6, victim quantity increases obviously as the number of nodes increases, but the number of messages is quite constant. Moreover, the total testing rounds increases slowly. This figure shows how the GT-based detection system efficiently operates when the number of nodes increases from $N = 500$ to $N = 1000$ with $m = 3$ and $J = 5$ jammers in a 1500×1500 network area. Therefore, the GT-based detection model is quite scalable.

Performance by the density of the network. We show how ITN reacts in the various network densities where the network field size broadens. With the given number of jammers and the increase of the network field, it is clear that the number of VN decreases where the system tries to deploy the nodes in order to cover the network field as much as possible. As we discussed before, due to the fact that the GT-based approach is disk-based classification of the nodes, sparse network would mainly help to reduce the number of VN, especially Δ, and then reduce the overall testing rounds as well. Figure 3.7 shows the various simulation results with the increasing network field size from 1500×1500 to 2500×2500 where $N = 1000$ with $m = 3$ and $J = 5$ jammers. As the network is sparse, the number of VN decreases as Δ gets smaller in this figure as we discussed.

Performance by the α in transmission range of the jammers. Since noise range is relatively larger than transmission range of sensor nodes, more messages of the sensor nodes will be jammed. In contrast, in the GT-based defense system, the larger jammers signal range may imply the increasing number of testing rounds even though this does not determine the damage area of the network. For instance, JAM [10] locks down the whole jammed region while the ITN algorithm minimizes the jammed region size by classifying the nodes more efficiency.

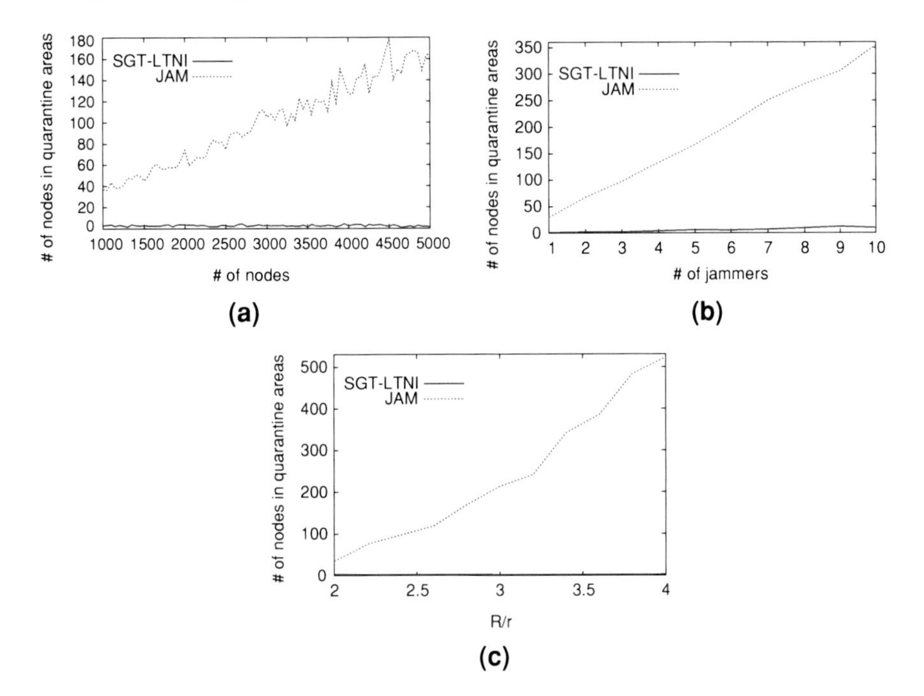

Fig. 3.9 Comparison of quarantine area with JAM algorithm

In Fig. 3.8, as the α gets larger, the number of VN increases since a jammer can transmit farther and contaminates more nodes during the activation. Moreover, more VN requires, more testing rounds to cull out the trigger nodes among them. In this result, the number of rounds rises very slowly as α gets larger.

Comparison of Performance to JAM approach. After constructing a jamming-resistant routing path by shifting all identified trigger nodes into receivers, there might be unreachable trigger nodes which are placed too deep to be reached by communication messages. The volume of unreachable trigger nodes could be compared to the number of jammed nodes from JAM algorithm [10] so as to determine the actual size of quarantine areas. The meaning of unreachable trigger nodes in ITN solution is the same to the jammed nodes in JAM algorithm, since the jammed nodes would not able to receive any messages according to [10], and unreachable trigger nodes also cannot receive any message either.

In Fig. 3.9, the size of unreachable trigger nodes is always substantially less then the size of jammed nodes from JAM algorithm. Especially, in Fig. 3.9a, only less than couple of nodes are unreachable trigger nodes and would not be able to receive messages where $n \in [500, 1000]$, whereas the number of jammed nodes becomes significantly larger due to higher network density.

As can be seen in Fig. 3.9b, the number of unreachable trigger nodes is less than 10, even in the presence of 10 jammers, but the number of jammed nodes sprouts with higher population with more jammers in WSNs.

As larger size of impact from jammers with bigger α, the jammed areas in the JAM approach gets inevitably bigger; however, ITN maintains a small number of trigger nodes, so that a significantly more nodes would be able to participate in secure communication than the JAM algorithm does. This fact clearly validates the efficiency and practicality of identification trigger nodes scheme.

References

1. Xu W, Ma K, Trappe W, Zhang Y (2006) Jamming sensor networks: attack and defense strategies. IEEE Network 20(3):41–47
2. Strasser M, Danev B, Capkun S (2010) Detection of reactive jamming in sensor networks. ACM Trans Sens Netw (TOSN) 7(2):1–29
3. Tague P, Nabar S, Ritcey JA, Poovendran R (2010) Jamming-aware traffic allocation for multiple-path routing using portfolio selection. IEEE/ACM Trans Networking 19(1):184–194
4. Mingyan Li, Koutsopoulos I, Poovendran R (2007) Optimal jamming attacks and network defense policies in wireless sensor networks. In: 26th, IEEE international conference on computer communications (INFOCOM), pp 1307–1315
5. Xu W, Trappe W, Zhang Y, Wood T (2005) The feasibility of launching and detecting jamming attacks in wireless networks. In: Proceedings of the 6th ACM international symposium on mobile ad hoc networking and computing (MobiHoc), pp 46–57
6. Cagalj M, Capkun S, Hubaux JP (2007) Wormhole-based antijamming techniques in sensor networks. IEEE Trans Mob Comput 6(1):100–114
7. Hang W, Zanji W, Jingbo G (2006) Performance of DSSS against repeater jamming. In: 13th IEEE international conference on electronics, circuits and systems (ICECS), pp 858–861
8. Sidek O, Yahya A (2008) Reed solomon coding for frequency hopping spread spectrum in jamming environment. Am J Appl Sci 5(10):1281–1284
9. Xu W, Wood T, Trappe W, Zhang Y (2004) Channel surfing and spatial retreats: defenses against wireless denial of service. ACM workshop on wireless security, pp 80–89
10. Wood AD, Stankovic J, Son S (2003) A jammed-area mapping service for sensor networks. In: 24th IEEE conference on real-time systems symposium (RTSS), pp 286–297
11. Shin I, Shen Y, Xuan Y, Thai MT, Znati T (2009) Reactive jamming attacks in multi-radio wireless sensor networks: an efficient mitigating measure by identifying trigger nodes. In: Proceedings of the 2nd ACM international workshop on foundations of wireless ad hoc and sensor networking and computing (FOWANC), pp 87–96
12. Shin I, Shen Y, Xuan Y, Thai MT, Znati T (2011) A novel approach against reactive jamming attacks, ad hoc and sensor wireless network, to appear ad hoc and sensor wireless networks 12(1–2):125–149
13. Bomze IM, Marco Budinich, Pardalos PM, Pelillo M (1999) Maximum clique problem, in handbook of combinatorial optimization. Kluwer Academic Publisher, Dordrecht, pp 1–74
14. Bron C, Kerbosch J (1973) Finding all cliques of an undirected graph. Commun ACM 16(9):575–577
15. Gupta R, Walrand J (2004) Approximating maximal cliques in ad-hoc networks, 15th IEEE international symposium on personal, indoor and mobile radio communications (PIMRC), vol 1. pp 365–369
16. Xuan Y, Shen Y, Nguyen NP, Thai MT (2011) A trigger identification service for defending reactive jammers in WSNs. IEEE Trans Mob Comput 99:, May 2011
17. Kaplan H, Katz M, Morgenstern G, Sharir M (2010) Optimal cover of points by disks in a simple polygon. In: Proceedings of the 18th annual European conference on algorithms:Part I (ESA), pp 475–486

Chapter 4
Randomized Fault Tolerant Group Testing and Advanced Security

Abstract In this chapter, we further optimize the time complexity of the scheme discussed in Chap. 3 and provide more advanced solutions. In detail, a randomized fault-tolerant group testing construction to reduce the computational cost, compared to the one using irreducible polynomials on Galois Field is introduced. Based on such a new GT construction, a more robust defense solution to an advanced jamming attack is presented for various network scenarios. Theoretical analysis and simulation results are included to validate the performance of this framework.

4.1 Advanced Attacker Model

In this chapter, we consider the same network model and attacker model as discussed in Chap. 3. However, we further consider a more advanced attacker model as follows:

Although the basic reactive jamming model is quite energy-efficient, the attackers may alter their behaviors to evade the detection, for which two advanced reactive jamming models: *probabilistic attack* and *asymmetric response time delay* are considered in this chapter. In the first one, the jammer responds to each sensed transmission with a probability η independently. In the second one, the jammer delays each of its jamming signals with an independently randomized time interval. In this chapter, we also handle the case $r \neq r_S$.

According to the jamming status, all the sensor nodes can be categorized into four classes: trigger nodes *TN*, victim nodes *VN*, boundary nodes *BN* and unaffected node *UN* as discussed in Chap. 3. Again, trigger nodes refer to the sensor nodes whose signals awake the jammers, i.e. within a distance less than r from a jammer. Victim nodes are those within a distance R from an activated jammer and disturbed by the jamming signals. Since $R > r$, $TN \subseteq VN$. Other than these disturbed sensors, *UN* and *BN* are the unaffected sensors while the latter ones have at least one neighbor in *VN*, hence $BN \subseteq UN$, and $VN \cap UN = \emptyset$. If a sensor belongs in two classes, it will label itself to the smallest one.

M. T. Thai, *Group Testing Theory in Network Security*, SpringerBriefs in Optimization, DOI: 10.1007/978-1-4614-0128-5_4,

4.2 Error-Tolerant Randomized Non-Adaptive Group Testing

As discussed in Chap. 1, a (*d,z*)-disjunct matrix has been used to handle the error tolerance. In the literature, numerous deterministic designs for (*d,z*)-disjunct matrix have been provided (summarized in [1]), however, these constructions often suffer from high computational complexity, e.g., calculation of irreducible polynomials on Galois Field, thus are not efficient for practical use and distributed implementation. In order to alleviate this testing overhead, the only randomized construction for (*d,z*)-disjunct matrix dues to Cheng's work via q-nary matrix has been proposed [2], which results in a (*d,z*)-disjunct matrix of size $t_1 \times n$ with probability p', where t_1 is

$$4.28\,d^2 \log \frac{2}{1-p'} + 4.28\,d^2 \log n + 9.84\,dz + 3.92\,z^2 \ln \frac{2n-1}{1-p'}$$

with time complexity $O(n^2 \log n)$.

Compared with this work, we present here a construction method which uses a classic randomized construction for d-disjunct matrix, namely, random incidence construction [3], to generate a (*d,z*)-disjunct matrix which can not only generate comparably smaller $t \times n$ matrix, but also handle the case where z is not known beforehand, instead, only the error probability of each test is bounded by some constant γ. Although z can be quite loosely upper bounded by γt, yet t is not an input. The motivation of this construction lies in the real test scenarios, the error probability of each test is unknown and asymmetric, hence it is almost impossible to evaluate z before knowing the number of pools.

4.2.1 *Construction of Randomized Error-Tolerant* (*d*, *z*)*-Disjunct Matrix*

The ETG algorithm as shown in Algorithm 10 is a randomized method to construct an error-tolerant (*d,z*)-disjunct matrix.

Algorithm 10 Error-Tolerant Group testing (ETG) construction

Algorithm 10 Error-Tolerant Group testing (ETG) construction

1: **Input**: n,d,z,s
2: **Output**: (*d,z*)-disjunct matrix with probability $(1-\frac{1}{s})$
3: Set $p = \frac{1}{d+1}$
4: Set $t = 2\left(\frac{(d+1)^{d+1}}{d^d}\right)(z - 1 + \ln s + (d+1)\ln n)$
5: Construct a $t \times n$ matrix M by letting each entry to be 1 with probability p
6: Return M

4.2.2 Theoretical Analysis

Theorem 4.1 *M is a (d,z)-disjunct matrix with* $t = 2\left(\frac{(d+1)^{d+1}}{d^d}\right)(z - 1 + \ln s + (d+1)\ln n)$ *rows with probability* $(1 - \frac{1}{s})$ *for a constant s where s can be arbitrarily large.*

Proof Note that M is not a *(d,z)*-disjunct matrix if for any single column c_0 and any other d columns $c_1, \dots c_d$, there are at most $z - 1$ rows where c_0 has 1 and all $c_1, \dots c_d$ have 0. By denoting $p = (\frac{1}{2})^l$, considering a particular column and d other columns in the matrix, the probability of such failure pattern is:

$$\sum_{i=0}^{z-1} \binom{t}{i} [p(1-p)^d]^i [1 - p(1-p)^d]^{t-i}$$

So use the union bound for all possible combinations and permutations of $(d+1)$ columns, we have the failure possibility bounded by

$$P_1 \le (d+1)\binom{n}{d+1} \sum_{i=0}^{z-1} \binom{t}{i} [p(1-p)^d]^i [1 - p(1-p)^d]^{t-i}$$

Here consider the CDF of binomial series and assume that $z - 1 \le tp(1-p)^d$ (**assert 1**), we then have

$$P_1 \le n^{d+1} exp\left(- \frac{(tp(1-p)^d - z + 1)^2}{2tp(1-p)^d} \right)$$

by *Chernoff bound*. To bound this by $\frac{1}{s}$, i.e.,

$$P_1 \le n^{d+1} exp\left(- \frac{(tp(1-p)^d - z + 1)^2}{2tp(1-p)^d} \right) \le \frac{1}{s}$$

we can derive that (**assert 2**)

$$p(1-p)^d \le \frac{z - 1 + \ln s + (d+1)\ln n}{t} - \frac{\sqrt{\ln^2(sn^{d+1}) + 2(z-1)\ln sn^{d+1}}}{t}$$

(infeasible by assert 1)
or

$$p(1-p)^d \ge \frac{z - 1 + \ln s + (d+1)\ln n}{t} + \frac{\sqrt{\ln^2(sn^{d+1}) + 2(z-1)\ln sn^{d+1}}}{t}$$

Therefore, we obtain the following lower bound

$$t \geq 2\left(\frac{(d+1)^{d+1}}{d^{d}}\right)(z - 1 + \ln s + (d+1)\ln n)$$

□

Corollary 4.1 *Given that each test has an independent error probability* γ, M is *(d,z)-disjunct matrix with* $t = \frac{\tau \ln n(d+1)^2 - 2\tau(d+1)\ln\frac{1}{s}}{(\tau-\gamma(d+1))^2}$ *with probability* $(1 - \frac{1}{s})$ *for arbitrarily large* s.

Proof Substituting z by γt in the above proof will complete this proof. □

Theorem 4.2 *The time complexity of the ETG algorithm is* $O(d^2 n \log n)$, *smaller than* $O(n^2 \log n)$, *provided that* $d < \sqrt{n}$.

Proof The proof is straightforward, thus omitted. □

4.3 Clique-Independent Set

Since identifying a set of interference-free testing teams using a disjoint disk-based method as discussed in Chap. 3 is time consuming, in this chapter, we present another method with using only local information to identify a set of interference-free testing teams. This method is based on the Clique-Independent Set (CIS).

Clique-Independent Set is the problem to find a set of maximum number of pair-wise vertex-disjoint maximal cliques [4]. Since this problem serves as the abstracted model of identifying interference-free testing teams, its hardness is of great interest in this scope. This problem has already been proved to be NP-hard for cocomparability, planar, line and total graphs, however its hardness on UDG is still an open issue, which we present its proof here.

4.3.1 NP-Completeness of CIS in UDGs

The NP-hardness of this problem on UDG is proven by reducing from the *Maximum Independent Set* problem on planar graph with maximum node degree 3 to it.

From [5], the *Maximum Independent Set* problem is NP-hard on planar graph with maximum degree 3, and from [6], any planar graph G with maximum degree 4 can be embedded in the plane using $O(|V|^2)$ area units such that its vertices are at integer coordinates and its edges consist of line segments of the form $x = i$ or $y = j$, for any integers i and j.

Theorem 4.3 *The CIS problem is NP-hard on Unit Disk Graph.*

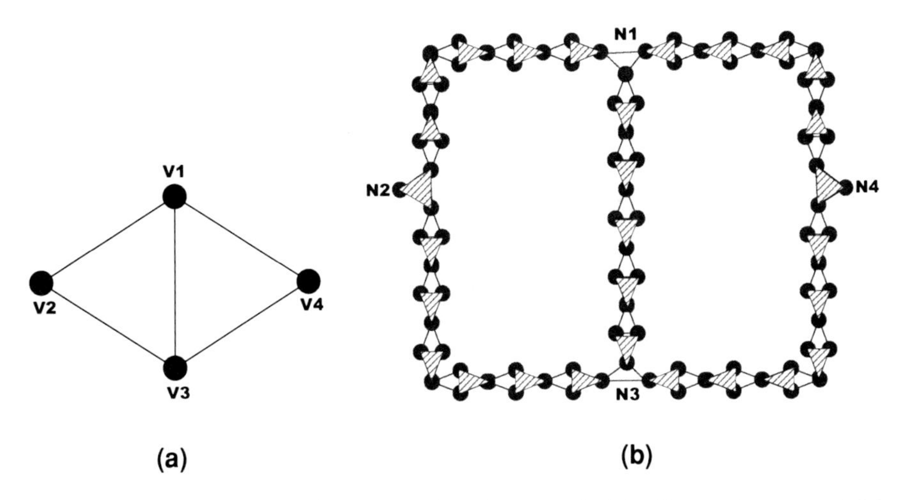

Fig. 4.1 Polynomial time reduction in the NP-hardness proof of the CIS problem on unit disk graphs. **a** $G' = (V', E')$. **b** $G = (V, E)$

Proof Given an instance $G' = (V', E')$ of such an *MIS* problem, whose optimal value is denoted as $MIS(G')$, we construct an instance $G = (V, E)$ of the *CIS* problem as follows:

- Embed G' in the plane in the way mentioned above [7].
- For each node $v_i \in V'$, attach two new nodes v_{i1} and v_{i2} to it and form a triangle $N_i = \{v_{i1}, v_{i2}, v_{i3}\}$, where each edge of this triangle N_i is of a unit length $r = \frac{\sqrt{3}}{3}$.
- Since each nodes v_i is incident to at most three edges, for all edges $(v_i, u), \dots, (v_i, v)$, move their endpoint from v_i to different $v_{ij}s$, e.g., (v_1, u) changes to (v_{11}, u) and (v_1, v) to (v_{12}, v). Afterwards for each of such edges $e = (u, v)$, assume that it is of length t, we divide it into t pieces and replace each piece with a concatenation of two triangles (not necessarily equilateral), as shown in Fig. 4.1b. Therefore, any edge $e_{ij} = (v_i, v_j) \in E'$ of length $|e_{ij}|$ becomes a concatenation of $2|e_{ij}|$ 3-cliques, denoted as $\{c_{ij}^{1,1}, c_{ij}^{1,2}, c_{ij}^{2,1}, \dots c_{ij}^{|e_{ij}|,1}, c_{ij}^{|e_{ij}|,2}\}$. Because of the triangles $N_i s$, the two triangles at each corner of Fig. 4.1b may need slight stenches, which can be done in polynomial time.
- The resulting graph G is then a unit disk graph with radius $r = \frac{\sqrt{3}}{3}$.

($\Rightarrow$): if G' has a maximum independent set M, for each $u_i \in M$, we choose cliques of two kinds in the corresponding instance G: (1) the clique N_i at u_i; (2) for each incident edge $e_{ij} = (u_i, u_j)$, choose cliques $\{c_{ij}^{1,2}, c_{ij}^{2,2}, c_{ij}^{3,2}, \dots, c_{ij}^{|e_{ij}|,2}\}$. Since the clique N_j at u_j shares a vertex with $c_{ij}^{|e_{ij}|,2}$, it cannot be selected. For any edge $e_{jk} = (u_j, u_k)$ where $u_j \notin M$ and $u_k \notin M$, choose cliques $\{c_{jk}^{1,2}, c_{jk}^{2,2}, \dots c_{jk}^{|e_{jk}|,2}\}$. It is easy to verify that all the cliques selected are vertex-disjoint from each other. Assume that after embedding G' into the plane, each node $v_i \in V'$ has coordinate

(x_i, y_i), then edge length $|e_{ij}| = \| v_i, v_j \|_1 = |x_i - x_j| + |y_i - y_j|$. Therefore if we have an independent set of size $|M| = k$ for G', we then have a clique-independent set of size $k' = k + \sum_{(i,j)\in E'} |e_{ij}|$.

($\Leftarrow$): if G has a clique-independent set of size k', since the lengths of the embedded edges are constant, then G' has exactly an independent set of size $k = k' - \sum_{(i,j)\in E'} |e_{ij}|$. The proof is complete. □

There have been numerous polynomial exact algorithms for solving this problem on graphs with specific topology, e.g., Helly circular-arc graph and strongly chordal graph [4], but none of these algorithms gives the solution on UDGs. In this chapter, we employ the *scanning disk approach* in [7] to find all maximal cliques on UDGs, and then find all the maximum CIS using a greedy algorithm.

4.4 Advanced Trigger Node Identification

We present a decentralized trigger identification procedure. It is lightweight in that all the calculations occur at the base station, and the transmission overhead as well as the time complexity is low and theoretically guaranteed. The algorithm still follows the main steps mentioned in Chap. 3. That is, we still use the same Node Classification procedure, the same Jamming Range Estimation procedure. After all victims are identified, they locally execute the Advanced Identification of Trigger Node (AITN) algorithm.

AITN has a similar concept to ITN in the way that they still divide victim nodes into many interference-free testing teams so that each testing team can be tested simultaneously using non-adaptive group testing. However, AITN uses a more sophisticated technique. Specifically, AITN uses Maximum Clique Independent Set (MCIS) to identify interference-free testing teams. And within each testing team, AITN uses randomized error-tolerant (d,z)-disjunct matrix to conduct the group testing, including an advanced method of estimating the upper bound d.

In this section, we only consider a basic attack model where the jammers *deterministically* and *immediately* broadcasts jamming signals once it senses the sensor signal. Therefore as long as at least one of the broadcasting victim nodes is a trigger, some jamming signals will be sensed, and vice versa. The performance of this protocol toward sophisticated attacker models with probabilistic attack strategies will be validated in the next section.

As shown in Table 4.1 , for each time slot, m sets of victim sensors will be tested. The selection of these sets involves a two-level grouping procedure.

First-level, the whole set of victims are divided into several interference-free testing teams. All the tests in different testing teams can be executed simultaneously since they will not interfere each other. Fig. 4.2 provides an example for this. 3 maximal cliques $C_1 = \{v_1, v_2, v_3, v_4\}$, $C_2 = \{v_3, v_4, v_5, v_6\}$, $C_3 = \{v_5, v_7, v_8, v_9\}$ can be found within 3 jammed areas. Imagine these three cliques are respectively the three teams we test at the same time. If v_4 in the middle team keeps broadcasting all

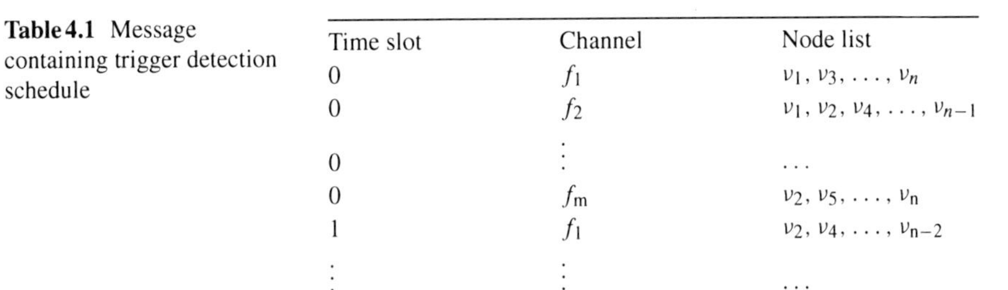

Table 4.1 Message containing trigger detection schedule

Time slot	Channel	Node list
0	f_1	$v_1, v_3, \ldots, v_n$
0	f_2	$v_1, v_2, v_4, \ldots, v_{n-1}$
0	$\vdots$	$\ldots$
0	f_m	$v_2, v_5, \ldots, v_n$
1	f_1	$v_2, v_4, \ldots, v_{n-2}$
$\vdots$	$\vdots$	$\ldots$

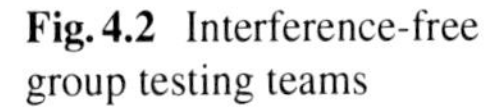

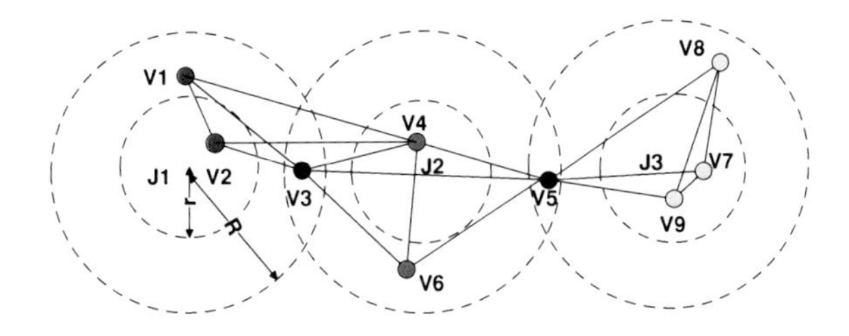

Fig. 4.2 Interference-free group testing teams

the time and J_2 is awaken frequently, no matter the trigger v_2 in the leftmost team is broadcasting or not, v_3 will always hear the jamming signals, so these two teams interfere each other. In addition, node-disjoint groups do not necessarily interference-free, as the leftmost and rightmost teams show.

Second-level, within each testing team, victims are further divided into multiple *testing groups*. This is completed by constructing a randomized $(d,1)$-disjunct matrix, as mentioned in Sect. 4.2, mapping each sensor node to a matrix column, and make each matrix row as a *testing group* (sensors corresponding to the columns with 1s in this row are chosen). Apparently tests within one group will possibly interfere that of another, so each group will be assigned with a different frequency channel.

The duration of the overall testing process is t time slots, where the length of each slot is $\mathscr{L}$. Both t and $\mathscr{L}$ are predefined, yet the former depends on the total number of victims and estimated number of trigger nodes, and the latter depends on the transmission rate of the channel. Specifically, at the beginning of each time slot, all the sensors designated to test in this slot broadcast a τ-bit test packet on the assigned channel to their 1-hop neighbors. Till the end of this slot, these sensors keep detecting possible jamming signals. Each sensors will label itself as a trigger unless in at least one slot of its testing, no jamming signal is sensed, in which case, the label is converted to a non-trigger.

4.4.1 Discovery of Interference-Free Testing Teams

As stated above, two disjoint sets of victim nodes are interference-free testing teams iff the transmission within one set will not invoke a jammer node, whose jamming signals will interfere the communications within the other set. Although we have estimated the jamming range R, it is still quite challenging to find these interference-free teams without knowing the accurate locations of the jammers. Notice that it is possible to discover the set of victim nodes within the same jammed area, i.e. with a distance R from the same jammer node. Any two nodes within the same jammed area should be at most $2R$ far from each other, i.e. if we induce a new graph $G' = (V', E')$ with all these victim nodes as the vertex set V' and $E' = \{(u, v) | \delta(u, v) \leq 2R\}$, the nodes jammed by the same jammer should form a clique. The maximum number of vertex-disjoint maximal cliques (i.e. CIS) of this kind provides an upperbound of possible jammers within the estimated jammed area, where each maximal clique is likely to correspond to the nodes jammed by the same jammer.

The solution consists of three steps: **CIS discovery** on the induced graph from the *remaining* victim without test schedules, boundary-based **local refinement** and interference -free **team detection**. We iterate three steps to decide the schedule for every victim node.

CIS discovery. As shown in Algorithm 11, this algorithm consists of two main steps: (1) employ Gupta's MCE algorithm [7] to find all the maximal cliques, and (2) use a greedy algorithm to obtain the CIS.

Algorithm 11 CIS discovery

1: **Input:** Induced Subgraph $G' = (W, E')$
2: **Output:** The set $\mathcal{C}$ of maximum number of disjoint maximal cliques.
3: Find out the set S of all maximal (not disjoint) cliques by using Gupta's *MCE* algorithm [7]
4: **while** $S \neq \emptyset$ **do**
5: Choose clique $C \in S$ which intersects with the *minimum* number of other cliques in S
6: $\mathcal{C} \leftarrow \mathcal{C} \cup \{C\}$
7: Remove all the *maximal* cliques intersecting with C
8: $S \leftarrow S \setminus \{C\}$
9: **end while**
10: return $\mathcal{C}$

Local Refinement. Each selected clique is expected to represent the jammed area attacked by the same jammer, and this area should not cover the boundary nodes. However, we did not take this into account when discovering the CIS, and need to locally update it. Specially, for each clique, we find its circumscribed circle CC and the concentric circle CC' with radius R of CC. In the case that CC' covers any boundary nodes, we locally select another clique by adding/removing nodes from this clique, to see if the problem can be solved. If not, we keep this clique as it is, otherwise, we update it. This is illustrated in Fig. 4.3 .

Team Detection. The cliques in CIS can also interfere each other, e.g. the clique $V_1 V_2 V_3 V_4$ and $V_5 V_7 V_8 V_9$ in Fig. 4.2 . This is because the signals from V_4 will wake J_2, who will try to block these signals with noises and affect V_5 by the way. But if any two cliques C_1 and C_2 are not connected by any single edge, then they are

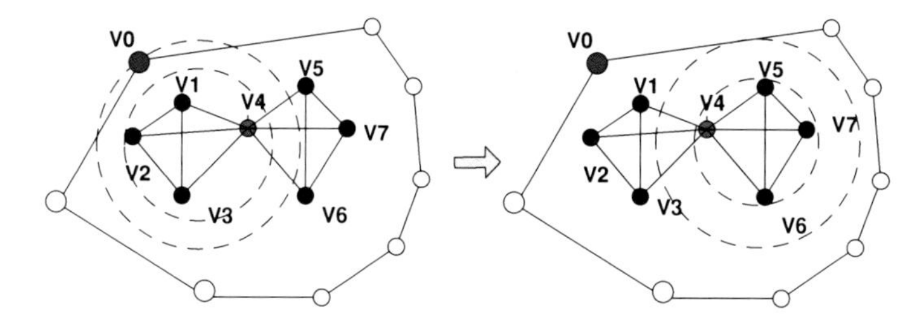

Fig. 4.3 clique $C_1 = V_1V_2V_3V_4$ is chosen by CIS, but its concentric circle CC' covers boundary node V_0, then clique $C_2 = V_4V_5V_6V_7$ replaces C_1 in the testing team for the first round. Clique $V_1V_2V_3$ are left for the next round.

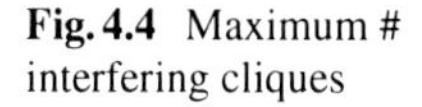
Fig. 4.4 Maximum # interfering cliques

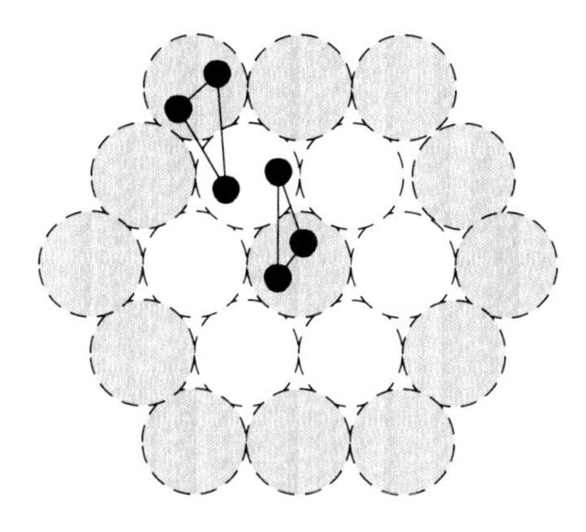

straightforwardly interference free, since the shortest distance between any node in C_1 and C_2 is larger than $2R$. But the farthest jammer waken by and from C_1 is $r < R$ distance away, whose jamming range can only reach another R distance further, which is thus away from C_2. Therefore, the cliques in the obtained CIS of this kind are selected as testing teams. While the others are left for the next time slot.

In addition, in the worst case, any single maximal clique C has at most 12 interfering cliques in the CIS, as the shadowed ones in Fig. 4.4. Therefore, at most 13 testing teams are required to cover all these cliques. If the number of channels k given is larger than 13, then a frequency division is available, i.e. these interfering cliques can still become simultaneous testing teams, on the condition each team can only use $\min\{\lceil \frac{k}{13} \rceil, m\}$ of the given channels, where m is the number of radios per sensor. Otherwise, we have to use time-divisions, i.e. they have to be tested in different time slots.

4.4.2 Estimation of Trigger Upperbound

Before bounding the trigger quantity from above, the triggering range r should be estimated. As mentioned in the attacker model, r depends not only on the power of both sensors and jammers, but also the jamming threshold θ and path-loss factor ξ:

$$r \geq \left(\frac{P_{\mathrm{n}} \cdot \theta}{P_{\mathrm{s}} \cdot Y}\right)^{\frac{1}{\xi}}$$

since the real time P_{n} and P_{s} are not given, r is estimated based on the SNR cutoff θ' of the network setting. In fact, the transmission range of each sensor r_{s} is a maximum radius to guarantee

$$\mathrm{SNR} = \frac{P_{\mathrm{a}}}{Pn} = \frac{P_{\mathrm{s}} \cdot Y}{P_{\mathrm{n}} \cdot r_{\mathrm{s}}^{\xi}} \geq \theta'$$

Therefore, we have:

$$r \approx r_{\mathrm{s}} \left(\frac{\theta}{\theta'}\right)^{\frac{1}{\xi}}$$

where θ' and ξ are parts of the network input, while θ is assumed as a constant, which indicates the aggressiveness of the jammer. For this estimation, θ can be first set as 10 db, which is the normally lower bound of SNR in wireless transmission, and then adaptively adjusted to polish the service quality.

With estimated r, since all the trigger nodes in the same team should be within a $2r$ distance from each other, by finding another induced graph $G'' = (W_{\mathrm{i}}, E'')$ from the victim nodes W_{i} in team i, with $E'' = \{(u, v) \in E'' \; if \; \delta(u, v) \leq 2r\}$, the size of the maximal clique indicates the upperbound of the trigger nodes, thus can be an estimate over d.

As mentioned above, all the parallel testing teams selected are interference free, therefore we roughly regard each team to be the jammed area of one jammer. As a deeper investigation, the number of jammers that can be invoked by the nodes in the same team (six 3-clique within the red circles) can be up to 6, since the minimum distance between two jammers is greater than R and $r \leq R$, as shown in Fig. 4.5. Therefore on the induced graph, the largest 6 cliques form the possible trigger set. However, since the jammer distribution cannot be that dense for the sake of energy-conserving, the former estimate over d is large enough.

4.4.3 Analysis of Time Complexity

By time complexity we mean the identification delay counted since the attack happens till all the nodes successfully identify themselves as trigger or non-trigger. Therefore,

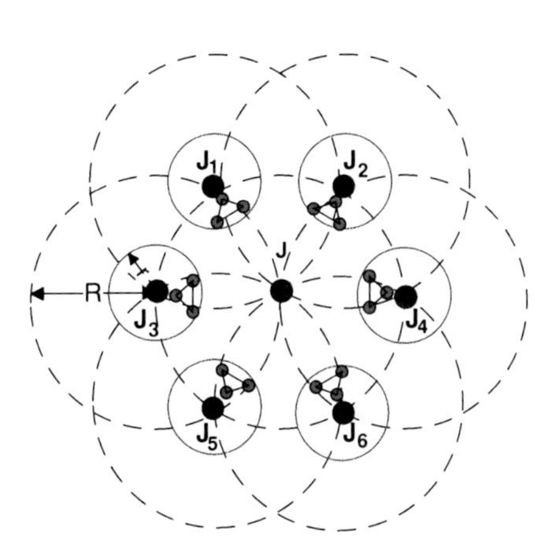

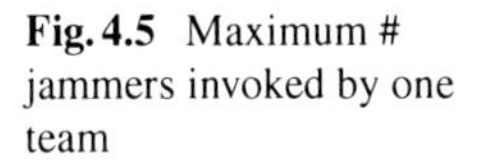

Fig. 4.5 Maximum # jammers invoked by one team

the complexity break downs into four parts: (1) the detection of jamming signals at local links T_d; (2) the routing of sensor report to the base station from each sensor node, and the testing schedule to each victim node from the base station, aggregated as T_r; (3) the calculation of CIS and R at the base station T_c; and (4) the testing at each jammed area T_t.

The local jamming signal detection involves the statistical properties of PDR, RSS and SNR, which is orthogonal to our work. We regard T_d as $O(1)$ since it is an entirely local operation and independent with the network scale.

The routing time overhead is quite complicated, since congestions need to be considered. For simplicity, we consider that all the 1-hop transmission takes $O(1)$ time and bound T_r using the diameter $\mathscr{D}$ of the graph. As mentioned earlier, the base station waits at most $O(2\mathscr{D})$ for the reports, so that is the upperbound of the one-way routing. As to the other way, we also bound it using $O(2\mathscr{D})$ to match any collision and retransmission cases.

The calculation of CIS resorts to the algorithm in [7], which finds $O(l\Delta)$ maximal cliques on UDG within $O(l\Delta^2)$ time, where $l = |E|$ and Δ refers to the maximum degree. We used a greedy algorithm to find a *MCIS* from these $O(l\Delta)$ cliques with $O(l^3\Delta^3\mathscr{Q})$ time: $O(l\Delta)$-time for each clique to check the overlapping with other cliques, $O(l\Delta)$-time to find a clique overlapping with minimum other cliques, and $\mathscr{Q}$ denotes the number of *testing teams*. Notice that in practice, sensor networks are not quite dense, so the number of edges l and maximum degree Δ are actually limited to small values. On the other hand, the time complexity of estimating R is up to $O(\frac{n\Delta}{2} + n(\log\frac{n\Delta}{2} + \log^6 n)$ using the minimum disk cover algorithm as mentioned.

The testing delay T_t depends on the number of testing rounds and the length of each round. Since the reactive jamming signal disappears as soon as these sensed 1-hop transmission finishes, each round length is then $O(1)$. The number of testing rounds is however complicated and bounded by Theorem 4.4.

Lemma 4.1. *Based on the ETG algorithm, the number of tests to identify d trigger nodes from* $|W|$ *victim nodes is upper bounded by* $t(|W|, d) = O(d_1^2\lceil\ln|W|\rceil)$ *w.h.p.*

Theorem 4.4 **(Main)** *The total number of testing rounds is upper bounded by*

$$O\left(\max_{i=1}^{\mathcal{Q}}\left\{\frac{13\min\{d_i^2\lceil \ln |W_i|\rceil, |W_i|\}}{m}\right\}\right)$$

w.h.p, with $d_i = \min\{\sum_{s=1}^{6} |c_s(G_i)|, |W_i|\}$ and $c_s(G_i)$ *is the* sth *largest clique over an induced unit disk subgraph* $G_i = (W_i, E_i, 2r)$ *in the testing team i.*

Proof First, from Lemma 4.1, at most $\frac{t(|W|,d)}{m} = \frac{d_i^2 \lceil \ln |W| \rceil}{m}$ testing rounds are needed to identify all nodes in *testing team i*. Second, the set of *testing teams* that can be tested in parallel is 13, as mentioned earlier. Combining with the worst-case upperbound of triggers in each team, the upperbound on round is derived. □

If the jamming range R is assumed known beforehand, similar to Chap. 3, the whole time complexity is thus

$$O\left(\max_{i=1}^{\mathcal{Q}}\left\{\frac{13\{d_i^2\lceil \ln |W_i|\rceil, |W_i|\}}{m}\right\}\right)$$

and asymptotically bounded by $O(n^2 \log n)$. It is asymptotically smaller than that of Chap. 3:

$$O\left(\sum_{i=1}^{\Delta(H)} \max_j \left[(1+o(1))\frac{d_j^2 \log_2^2 |W_j|}{\log_2^2(d_j \log_2 |W_j|)}/m\right]\right)$$

where $\Delta(H)$ refers to the maximum degree of the induced graph H (in this new solution, maximum degree is not involved). By taking the calculation overhead for R into account, the overall time complexity is asymptotically $O(n^2 \log n + n \log^6 n)$, which is $O(n \log^6 n)$ for $n \geq 4$.

4.5 Advanced Solutions Toward Sophisticated Attack Models

In this section, we consider two sophisticated attacker models: *probabilistic attack* and *variant response time delay*, where the jammers rely each sensed transmission with different probabilities, instead of deterministically, or delay the jamming signals with a random time interval, instead of immediately. This advanced attack will bring some testing errors in the AITN algorithm. Thus we will utilize the error-tolerant disjunct matrix and relax the identification procedures to asynchronous manner to handle these cases. Some notations can be found in Table 4.2 . In this section, the terms *test* and *group*, the terms *column* and *nodes* are interchangeable.

Table 4.2 Notations

Notation	Content
T^+	The number of false positive outcomes
T^-	The number of false negative outcomes
$u(i)$	The number of trigger nodes in test i
$x(i)$	The reaction time of jammer toward test i
$g(i)$	The outcome of test i

4.5.1 Upper Bound on the Expected Value of z

We investigate the properties of both jamming behaviors and obtain the expected number of error tests in both cases of jamming models through the following analysis. Since in practice, it is not trivial to establish accurate jamming models, we derive an upper bound of the error probability which does not require the beforehand knowledge of the objective jamming models, which is therefore feasible for real-time identifications. Since it is a relaxed bound, it could be further strengthened via learning the jamming history.

4.5.1.1 Probabilistic Jamming Response (Detection)

In this model, a jammer can choose not to respond to some sensed ongoing transmissions, in order to evade the detection. Assume that each ongoing transmission has an independent probability η to be responded. In the ETG construction, where each matrix entry is IID and has a probability p to be 1, therefore for any single test i with $i \in [1, t]$:

$$\mathbf{Pr}[u(i) = x] = \binom{d}{x} p^{\mathrm{x}}(1-p)^{\mathrm{d-x}} \tag{4.1}$$

Hence for each test i, the event that it contains at least one trigger but returns a negative result has a probability at most:

$$\mathbf{Pr}[g(i) = 0 \,\&\, u(i) \geq 1] \tag{4.2}$$

$$= \sum_{x=1}^{\mathrm{d}} (1-\eta)^{\mathrm{x}} \binom{d}{x} p^{\mathrm{x}}(1-p)^{\mathrm{d-x}} \tag{4.3}$$

$$= [(1-\eta)p + 1 - p]^{\mathrm{d}} - (1-p)^{\mathrm{d}} \tag{4.4}$$

$$= (1-\eta p)^{\mathrm{d}} - (1-p)^{\mathrm{d}} < (1-\eta)p \tag{4.5}$$

Meanwhile, the event that it contains no trigger nodes but returns a positive result, has a probability:

$$\mathbf{Pr}[g(i) = 1 \,\&\, u(i) = 0] = 0 \tag{4.6}$$

Since in practical $\eta \geq \frac{1}{2}$, we therefore have the expected number of false positive and negative tests is respectively at most $pt/2$ and *0*.

Instead of the jamming behavior, the jamming signal detection errors can be analyzed using the same method. Given that each node detects possible jamming signals successfully with probability q, then following Eq. 4.1, we can similarly have the false negative rate of each test i:

$$\mathbf{Pr}[g(i) = 0 \,\&\, u(i) \geq 1] \tag{4.7}$$

$$= \sum_{x=1}^{\mathrm{d}} (1-q)^{\mathrm{x}} \binom{d}{x} p^{\mathrm{x}} (1-p)^{\mathrm{d-x}} \tag{4.8}$$

$$= [(1-q)p + 1 - p]^{\mathrm{d}} - (1-p)^{\mathrm{d}} \tag{4.9}$$

$$= (1-qp)^{\mathrm{d}} - (1-p)^{\mathrm{d}} < (1-q)p \tag{4.10}$$

which is also small considering $p = \frac{1}{d+1}$.

4.5.1.2 Variant Reaction Time

In this attack model, a jammer would not respond intermediately after sensing the ongoing transmissions, but instead wait for a randomized time delay, thereby generating errors in the test outcomes. Since it is expensive to synchronize the tests among sensors, we use a predefined testing length as $\mathscr{L}$, thus the test outcome of test $i \in [1, t]$ is generated within time interval $[(\lceil \frac{i}{m} \rceil - 1)\mathscr{L}, \lceil \frac{i}{m} \rceil \mathscr{L}]$. For any test i, there are two possible error events as follows:

- *Fp*(i): test i is negative, but some jamming signals are delayed from previous tests and interfere this test, where we have a false positive event;
- *Fn*(i): test i is positive, but the jammer activated in this test delayed its jamming signals to some subsequent tests, meanwhile, no delayed jamming signals from previous tests exists, where we have a false negative event.

Since the jammers are assumed to block communications only on the channels where transmissions are sensed, for the following analysis, we claim that the interferences can only happen between any two tests i,j with $i \equiv j (\mathbf{mod}\ m)$. Denote the delay of jamming signals as a random variable $X = \{x(1), x(2), x(3), \ldots x(t)\}$ where $x(i)$ is the delay for possible jamming signals arisen from test i. (1) For event *Fp*(i), consider the test $i - m$, in order to have its jamming signals delayed to test i, we have a bound on $x(i - m) \in (0, 2\mathscr{L})$. Similarly, in order to have the signals of any test j delayed to i, we have $x(j) \in [(\frac{i-j}{m} - 1)\mathscr{L}, (\frac{i-j}{m} + 1)\mathscr{L}]$. Further the probability density function of X is $\mathscr{P}(i) = \mathbf{Pr}[X = x(i)]$. Consider all the tests prior to i, which are i **mod** m, $1 + i$ **mod** $m, \ldots, i - m$, we then have the probability for $Fp(i)$:

$$(1-p)^{\mathrm{d}} \sum_{\mathrm{j=i\ \mathbf{mod}\ m}}^{\mathrm{i-m}} \int_{\left(\frac{\mathrm{i-j}}{\mathrm{m}}-1\right)\mathscr{L}}^{\left(\frac{\mathrm{i-j}}{\mathrm{m}}+1\right)\mathscr{L}} \mathscr{P}(w)\mathbf{d}w(1-(1-p)^{\mathrm{d}}) \tag{4.11}$$

To simplify this expression, we assume that $X/\mathscr{L}$ follows a uniform distribution within the range $[0, \beta]$ with a small β, which is reasonable and efficient for attackers in practice. Since the nature of jamming attacks lies in adapting the attack frequency due to the sensed transmissions, too large delay does not make sense to tackle the ongoing transmissions. Under a uniform distribution, the probability of $Fp(i)$ becomes:

$$\begin{aligned}
&(1-(1-p)^{\mathrm{d}})(1-p)^{\mathrm{d}} \sum_{\mathrm{j=max\,i\ \mathbf{mod}\ m, i-m-\beta-1}}^{\mathrm{i-m}} \frac{2}{\beta} \\
=&(1-(1-p)^{\mathrm{d}})(1-p)^{\mathrm{d}} \left(\left\lceil \frac{i}{m} \right\rceil - 1\right)\frac{2}{\beta}
\end{aligned}$$

Therefore, the expected number of false positive tests is at most

$$\begin{aligned}
T^{+} \leq& \sum_{\mathrm{i=1}}^{\mathrm{t}} (1-(1-p)^{\mathrm{d}})(1-p)^{\mathrm{d}}(\beta)\frac{2}{\beta} \\
\leq& 2\sum_{\mathrm{i=1}}^{\mathrm{t}} (1-(1-p)^{\mathrm{d}})(1-p)^{\mathrm{d}} \\
\leq& 2(1-(1-p)^{\mathrm{d}})(1-p)^{\mathrm{d}} t
\end{aligned}$$

(2) For event $Fn(i)$, following the similar arguments above, we have an upper bound of the probability for $Fn(i)$ (assume that any delays larger than l at test i will interfere the tests j following i where $j \in [\max(i\,\mathbf{mod}\,m, i - m - \beta - 1), i - m])$:

$$\begin{aligned}
&(1-(1-p)^{\mathrm{d}}) \int_{1}^{+\infty} \mathscr{P}(w)\mathbf{d}w \\
&\cdot \left(1 - \sum_{j} \int_{\left(\frac{i-j}{\mathrm{m}}-1\right)\mathscr{L}}^{\left(\frac{i-j}{\mathrm{m}}+1\right)\mathscr{L}} \mathscr{P}(w)\mathbf{d}w(1-(1-p)^{\mathrm{d}})\right) \\
\leq&(1-(1-p)^{\mathrm{d}})(1-2(1-(1-p)^{\mathrm{d}}))(\beta - l)/\beta \\
\leq&(1-(1-p)^{\mathrm{d}})(1-2(1-(1-p)^{\mathrm{d}}))
\end{aligned}$$

So the expected number of false negative tests is at most

$$T^{-} \leq (1-(1-p)^{\mathrm{d}})(1-2(1-(1-p)^{\mathrm{d}}))t \quad (4.12)$$

Therefore, we could use a union bound and obtain a worst-case error rate of each test:

$$\begin{aligned}\gamma =& \frac{p}{2} + 2(1-(1-p)^{\mathrm{d}})(1-p)^{\mathrm{d}} \\ &+ (1-(1-p)^{\mathrm{d}})(1-2(1-(1-p)^{\mathrm{d}})) \\ =& (10\tau - 8\tau^2 - \tau^{-\mathrm{d}} - 1)/2\end{aligned}$$

where $\tau = (d/(d+1))^{\mathrm{d}}$.

Intuitively, we can have an upper bound on the number of error tests as $z = \gamma t = (10\tau - 8\tau^2 - \tau^{-\mathrm{d}} - 1)/2$, and take it as an input to construct the (d,z)-disjunct matrix. However, notice that z depends on t, i.e., the number of rows of the constructed matrix, we therefore derive another bound of t related to γ, as shown by Corollary 4.1.

4.5.2 *Error-Tolerant Asynchronous Testing Within Each Testing Team*

By applying the derived worst-cast number of error tests into the ETG construction, we can obtain the following algorithm where tests are conducted in an asynchronous manner to enhance the efficiency.

As shown in Algorithm 12, after all the groups are decided, conduct group testing on them in m pipelines, where in each pipeline any detected jamming signals will end the current test and trigger the next tests while groups receiving no jamming signals will be required to resend triggering messages and wait till the predefined round time has passed. These changes over the original algorithm, especially the asynchronous testing are located in each testing team, thus will not introduce significant overheads, however, the resulted error rates are limited to a quite low level.

Algorithm 12 Asynchronous Testing

1: **Input:** n victim nodes in a *testing team*
2: **Output:** all trigger nodes within these victim nodes
3: Estimate d as mentioned {upper bound of error probability for each test}
4: Set $\gamma = (10\tau - 8\tau^2 - \tau^{-\mathrm{d}} - 1)/2$
5: Set $t = \frac{\tau \ln n (d+1)^2}{(\tau - \gamma(d+1))^2}$
6: Construct a (d,z)-disjunct matrix using ETG algorithm with t rows, and divide all the n victim nodes into t groups accordingly $\{g_1, g_2, \cdots, g_t\}$
{For each round, conduct group testing on m groups using m different channels

(radios). The testing is asynchronous in that, the m groups tested in parallel do not wait for each other to finish the testing, instead, any finished test j will trigger the test $j + m$, i.e., the tests are conducted in m pipelines. }

7: **for** $i = 1$ To $\lceil t/m \rceil$ **do**

8: Conduct group testing in groups $g_{\mathrm{im+1}}$, $g_{\mathrm{im+2}}$, $g_{\mathrm{im+m}}$ in parallel

9: If any nodes in group g_{j} with $j \in [im + 1, im + m]$ detects jamming noises, the testing in this group finishes and start testing on $g_{\mathrm{j+m}}$

10: If no nodes in group g_{j} detect jamming noises, while at least one other test in parallel detects jamming noises, let all the nodes in group g_{j} resend 3 more messages to activate possible hidden jammers.

11: If no jamming signals are detected till the end of the predefined round length ($\mathscr{L}$), return a negative outcome for this group and start testing on $g_{\mathrm{j+m}}$

12: **end for**

4.6 Experimental Analysis

4.6.1 Overview

In this section, we present an evaluation of this advanced trigger identification from the following three facets:

- in order to show the benefit of this detection scheme, a simulation was conducted to compare it with JAM [8] in terms of the end-to-end delay and delivery ratio of the detour routes from the base station to all the sensor nodes, as the number of sensors n, sensor range r_s and number of jammers J vary within practical intervals.
- in order to show the acceleration effect of the CIS in this solution, we compare the complexity of this solution to the disjoint disks-based method mentioned in Chap. 3, with varying the above four parameters, where both jamming and triggering range R and r are assumed to be known beforehand.
- in order to show its performance and robustness toward advanced attackers, we provide its false positive/negative rate, when taking into account those two advanced jammer models, as well as the estimation of R.

The simulation is developed using C++ on a Linux Workstation with 8GB RAM. A $1,000 \times 1,000$ square sensor field is created with uniformly distributed n sensor nodes, one base station and J randomly distributed jammer nodes. All the simulation results are derived by averaging 20 random instances.

4.6.2 Benefits for Jamming-Resistent Routing

JAM [8] proposed a jamming-resistent routing scheme, where all the detected jammed areas will be evaded and packets will not pass through the jammed nodes.

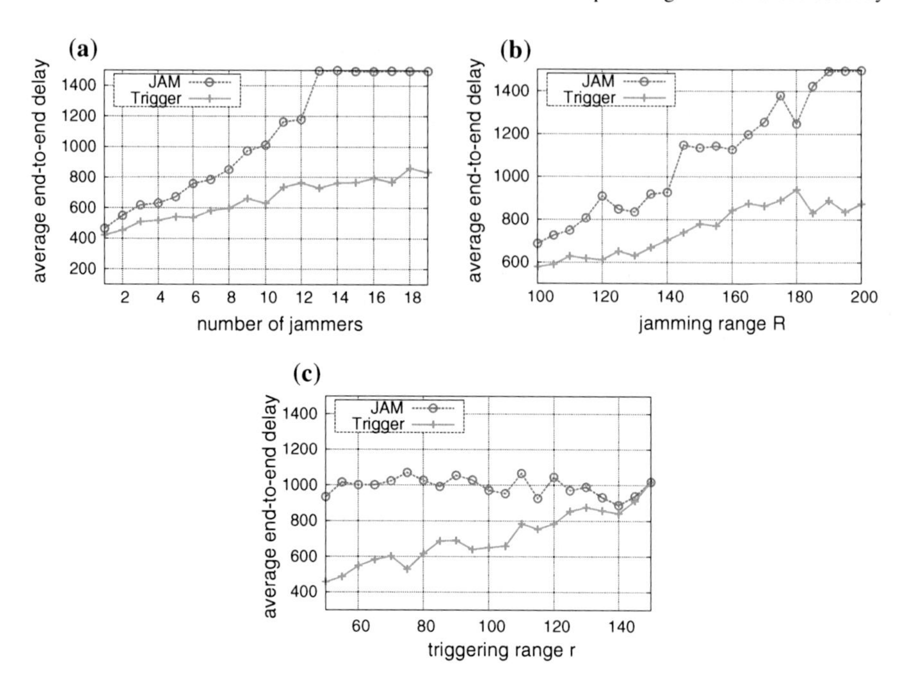

Fig. 4.6 Benefits for routing. **a** Average end-to-end delay by J. **b** Average end-to-end delay by R. **c** Average end-to-end delay by θ

We compare the end-to-end delay between each sensor node and the base station, of the selected routes by evading the jammed areas detected by JAM, with that of the ones evading only trigger nodes.

Three key parameters for routing could be the number of Jammers J, jamming range R, jamming threshold θ. As mentioned earlier, θ indicates the aggressiveness of the attacker and the triggering range $r \approx r_s(\frac{\theta}{\theta'})^{\frac{1}{\xi}}$. Therefore, with r_s, θ' and ξ as fixed network inputs, the effect of θ can be exactly indicated by studying the effect of r instead.

The whole network has $n = 1,500$ nodes and sensor transmission range $r_s = 50$. The results with respect to the three parameters $J \in [1, 20]$, $R \in [100, 200]$, $r \in [50, 150]$ are included in Fig. 4.6a, b and c respectively. Notice that for each experiments, the other two parameters are set as the median value of their corresponding intervals. Therefore, $R = 150$ for Fig. 4.6c, which matches the extreme case $R = r$. Furthermore, for the nodes that are in jammed areas for JAM and that are triggers for our method, in another word, unable to deliver packets to or from the base station, we count the delay as $n + 1$, which is an upperbound of the route length.

As shown in Fig. 4.6a and b, when j and R increases, the routing delay goes up, which is quite reasonable since the jamming areas get larger and more detours have to be taken. The length of routes based on JAM quickly climbs up to the upperbound, while that of our trigger method is much lower and more stable, specifically keeps less than 900 s. When triggering range r is small, as in Fig. 4.6c, the end-to-end delay of Trigger-based routing is much smaller than the other, while as r increases the two approaches each other, since more victim nodes are triggers now.

4.6.3 Improvements on Time Complexity

As discussed earlier, a disk-based solution to identify interference-free testing teams in Chap. 3 has a high time complexity due to its centralized algorithm. Instead, an independent clique-based method presented in this chapter is proved to be asymptotically lower than the previous, while the message complexities are approaching each other. In this section, we verify this bound through simulations on network instances with various settings. Specifically, the network size n ranging from 450 to 550 with step 2, transmission r_s from 50 to 60 with step 0.2 and number of jammers J from 3 to 10 with step 1. Parameter values lower than these intervals would make the sensor network less connected and jamming attack less severe, while higher values would lead to impractical dense scenarios and unnecessary energy waste.

Since the length of each reactive attack is equal to the transmission delay of the object sensor signal, note that in the advanced trigger detection, only one message is broadcast by each sensor in the testing groups. Therefore, it is reasonable to predefine the length of each testing round as a constant. We set this as 1 s, which is far more enough for any single packet to be transmitted from one node to its neighboring nodes. Henceforth, the time cost shown in Fig. 4.6 only indicates the number of necessary rounds to identify all the triggers, and can be further reduced. The message complexity is measured via the average message cost on each sensor node.

As shown in Fig. 4.7a and b, this clique-based scheme completes the identification with steadily less than 10 s, compared to the increasing time overhead with more than 15 s of the disk-based solution, as the network grows denser with more sensor nodes. Meanwhile, its amortized communication overheads are only slightly higher than that of the other solution, whereas both are below 10 messages per victim node. Therefore, the independent clique-based scheme is more efficient and robust to large-scale network scenarios.

With the sensor transmission radius growing up, the time complexity of the disk-based solution gradually ascends (Fig. 4.7c and d) due to the increased maximum degree $\Delta(H)$ mentioned in the above analysis. Comparatively, the time cost of clique-based solution remains below 10 s, while the message complexity still approximates the other one.

Since sensor nodes are uniformly distributed, the more jammer nodes placed in the networks, the more victim nodes are expected to be tested, the identification complexity will therewith raises, as the performance of disk-based scheme shows in Fig. 4.7e and f. Encouragingly, the independent clique based scheme can still finish the identification promptly with less than 10 s, which grows up much slower than the

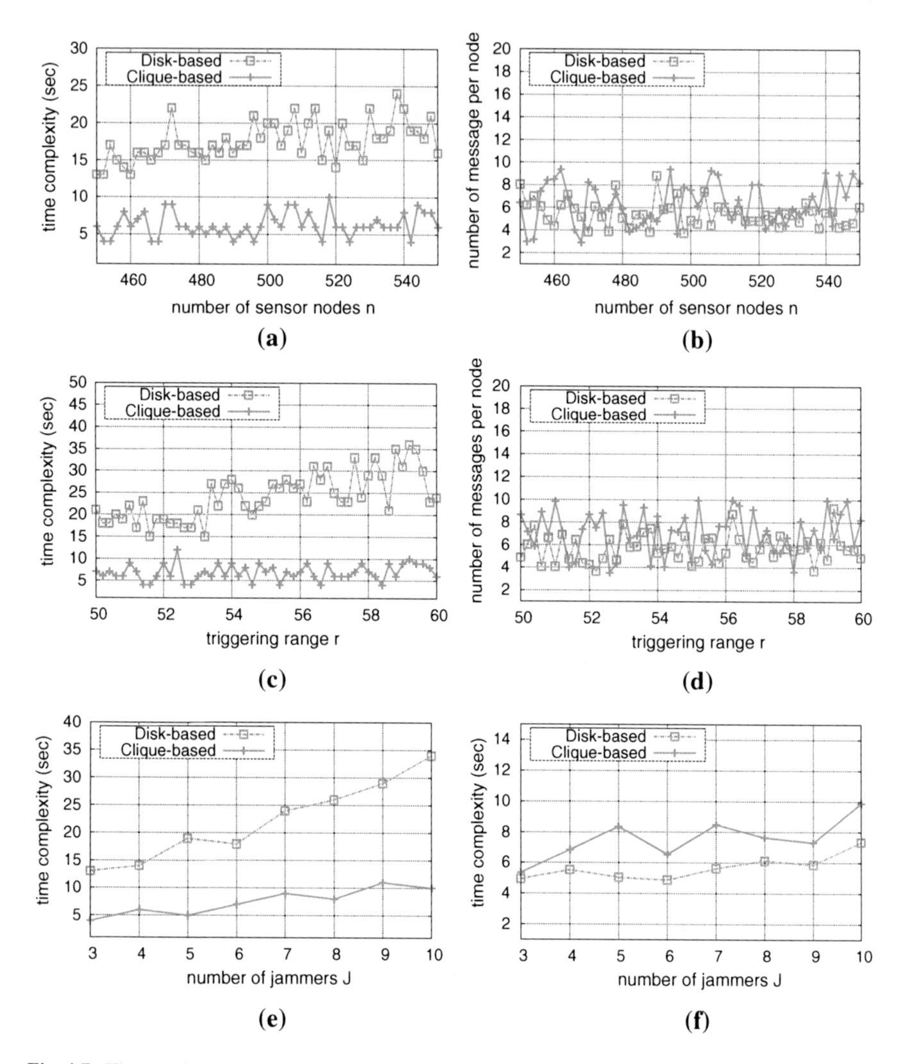

Fig. 4.7 Time and message complexity. **a** # rounds by n. **b** # messages by n. **c** # rounds by r. **d** # messages by r. **e** # rounds by J. **f** # messages by J

other. It has slightly more communication overheads (10 messages per victim nodes) but is still affordable to power-limited sensor nodes.

4.6.4 Robustness to Various Jammer Models

Finally, we present the robustness of this advanced trigger identification under different jamming environments by varying the two parameters of the jammer behaviors above: *Jammer Response Probability* η and *Testing Round Length/Maximum*

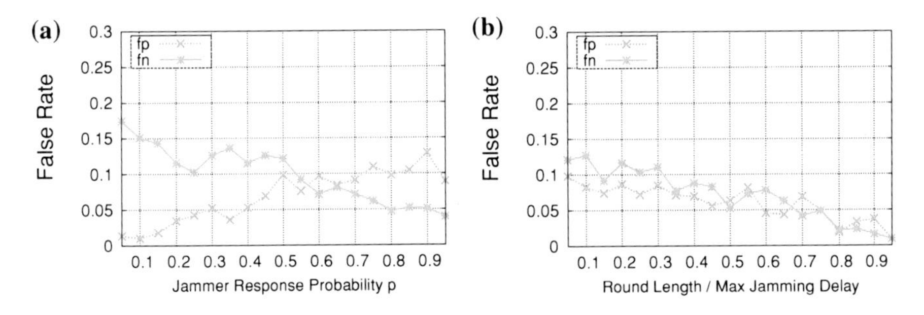

Fig. 4.8 Solution robustness. **a** Probabilistic jammer response. **b** Random jamming delay

Jamming Delay $\mathscr{L}/X$ and illustrate the resulted false rates in Fig. 4.8a and b. To simulate the most dangerous case, a hybrid behavior for all the jammers are investigated. For example, the jammers in the simulation of Fig. 4.8a not only launch the jamming signals probabilistically, but also delay the jamming messages with a random period of time up to $2\mathscr{L}$. On the other hand, the jammers in the simulation of Fig. 4.8b respond each sensed transmission with probability 0.5 as well. All the simulation results are derived by averaging 10 instances for each parameter team.

As shown in both figures, the *extreme* cases where jammers respond transmission signals with a probability as small as 0.1, or delay the signals to up to 10 testing rounds later are considered. This actually contradicts with the nature of reactive jamming attacks, which aim at disrupting the network communication as soon as any legitimate transmission starts. The motivation of such parameter setting is to show the robustness of this scheme even if the attackers sense the detection and intentionally slow down the attacks. The overall false rates are below 20% for any parameter values.

In Fig. 4.8a, when $\eta > 1/2$ which corresponds to practical cases, the false negative rates generally decrease from 10 to 5% as η increases. Meanwhile the false positive rate grows gently, but is still below 14%, this is because as more and more jamming signals are sent, due to their randomized time delays, more and more following tests will be influenced and become false positive. In Fig. 4.8b, considering the practical cases where $\mathscr{L}/X > 1/2$, both rates are going down from around 10 to 1%, since the maximum jamming delay becomes shorter and shorter compared to the testing round length $\mathscr{L}$, in which case, the number of interferences between consecutive tests is decreasing.

References

1. Du DZ, Hwang FK (2006) Pooling designs: group testing in molecular biology. World Scientific, Singapore
2. Xu W, Ma K, Trappe W, Zhang Y (2006) Jamming sensor networks: attack and defense strategies. IEEE Network 20:41–47

3. Dyachkov AG, Nabar S, Rykov VV, Rashad AM (1989) Superimposed distance codes. Prob Control Inf Theor 18(4):237–250
4. Sekar V, Duffield N, van der Merwe K, Spatscheck O, Zhang H (2006) LADS: Large-scale automated DDoS detection system. In: Proceedings of the annual conference on USENIX'06 annual technical conference, Boston, p 16
5. Xu W, Wood T, Trappe W, Zhang Y (2004) Channel surfing and spatial retreats: defenses against wireless denial of service. ACM workshop on wireless security, pp 80–89
6. Mingyan Li, Koutsopoulos I, Poovendran R (2007) Optimal jamming attacks and network defense policies in wireless sensor networks. In: Proceedings of the 26th IEEE international conference on computer communications (INFOCOM'07), pp 1307–1315
7. Khattab SM, Sangpachatanaruk C, Mosse D, Melhem R, Znati T (2004) Honeypots for mitigating service-level denial-of-service attacks. In: Proceedings of the 24th IEEE international conference on distributed computing systems (ICDCS'04), pp 328–337
8. Tague P, Nabar S, Ritcey JA, Poovendran R (2010) Jamming-aware traffic allocation for multiple-path routing using portfolio selection. IEEE/ACM Trans Networking 19(1):184–194

Chapter 5
Outlooks

Abstract In this concluding chapter, the outlook and open group testing-related problems in network security are presented. In particular, we discuss about a low time complexity for matrix construction, fault-tolerant group testing construction, and the theory aspect of size constraint group testing as well as applications of the trigger nodes detection approach.

5.1 General Detection Framework Based on Group Testing

This book has brought our attention to an emergent topic of using the group testing theory for several advanced network security problems. As can be seen in Chap. 2, GT based solution has provided an advanced detection framework for the detecting application DoS attack with a low detection latency and near-zero false positive/negative rate. With this promising result, one may be encouraged to investigate this underlying theoretical framework for the detection against a general case of network assaults, where malicious requests are indistinguishable from normal ones. At first, one may want to enhance the detection efficiency of this framework and further lower the false negative/positive rate. Some possible directions for this can be: (1) the sequential algorithm can be adjusted to avoid the requirement of isolating attackers; (2) investigate more efficient d-disjunct matrix which could dramatically decrease the detection latency; (3) the overhead of maintaining the state transfer among virtual servers can be further decreased by more sophisticated techniques; and (4) although the current framework already has quite low false positive/negative rate, one can still improve it via error-tolerant group testing methods, using the one presented in Chap. 4 or others [1].

One immediate application of our framework is protocol-layer attacks—SYN flood [2] where victim servers are exhausted by massive half-open connections. Although these attacks occur in different layers and are of different styles, the victim machines will gradually run out of service resource and indicate anomaly. Since

M. T. Thai, *Group Testing Theory in Network Security*, SpringerBriefs in Optimization,
DOI: 10.1007/978-1-4614-0128-5_5, © My T. Thai 2012

the framework discussed in Chap. 2 mechanism only relies on the feedback of the victims, instead of monitoring the client behaviors or properties, it is promising to tackle these attack types.

5.2 Size Constraint Group Testing

There are several open problems for this variant. In this book, we have discussed the sequential group testing for the size constraint. However, we have not investigated the sufficient and necessary conditions for a non-adaptive size constraint group testing.

As mentioned, the main difference between this model and other existing GT models is that it has a size constraint on both the maximum number of items grouped together and the maximum number of pools used. With these size constraints, it poses several challenges in the group testing theory and requires an in-depth analysis of the matrix construction complexity.

Consider a binary matrix $M_{t\times n}$ and the outcome vector V. Let a row weight $\lambda_i = \sum_{j=1}^{n} M[i, j]$, a column weight $\omega_j = \sum_{i=1}^{t} M[i, j]$ and $f(\lambda_i, \omega_j, r_{\text{normal}})$ be a function of server load based on the number of active clients; thus $f(\lambda_i, \omega_j, r_{\text{normal}}) \leq w$ where w is the server capacity. We may want to investigate the following problem:

Size Constraint Matrix Construction Given positive integers $n, d \ll n, w$ and the function f, construct a d-disjunct matrix $M_{t\times n}$ so as to minimize the number of rows t and $f(\lambda_i, \omega_j, r_{normal}) \leq w$.

Some viable directions to address the above problem are as follows:

- Study the necessary conditions for such a d-disjunct matrix to exist, that is, study the asymptotic lower and upper bounds on t with respect to d, n, f, w. Design algorithms to construct M
- Further investigate into an error-tolerant of M, that is, construct a (d, z)-disjunct matrix satisfying the above conditions.

5.3 Jamming Attacks and Trigger Node Detection

As evident in Chaps. 3 and 4, the trigger node detection concept provides an excellent way to handle the jamming attacks. Another benefit of identifying the trigger nodes is to help construct a routing protocol which does not activate any reactive jammer. That is, trigger nodes should be receivers only in the routing.

We present a simple routing algorithm based on the Connected Dominating Set (CDS), which is defined as: Given a graph $G=(V,E)$, find a subset $C \subseteq V$ with the minimum size such as for any vertex $v \in V$, v is either in C or adjacent to at least one vertex in C and the induced graph on C is connected. As is well known in the literature, CDS is one of the efficient ways of constructing a routing protocol where nodes in C, called backbone, handle all the transmission. Therefore, the smaller the C, the better the routing method.

Consider network $G = (V,E)$ with $U \subset V$ as a set of identified trigger nodes. A directed graph $G' = (V, E')$ is constructed by changing all the undirected edges $(u, v) \in E$, where $u \in V \setminus U$ and $v \in U$ to the directed edge (u, v). We then deploy a CDS algorithm in directed graph [3] on G'. It is easy to see that the obtained CDS C will not consist of any node in U. Finally, we construct a broadcast tree T by connecting nodes in C to the rest using newly added directed edges.

Another viable way is to design a jamming-resistant routing, coupled with the jamming noise detection during the construction of jamming-resistant routing so that we can route "through" the jammed areas.

One interesting research topic based on the trigger nodes identification is the jammer localizations. Given a set of identified triggers and their locations, we may be able to use the current localization methods to locate the jammers.

Another open problem along this work is the jammer mobility. Although the identification latency has been shown small, it would not be efficient toward jammers that are moving at a high speed.

From the security point of view, some problems may require the set of items in each testing group to be able to communicate to each other. That is, consider graph $G=(V, E)$ representing a network where V is a set of nodes and E is the set of communication links. $(u, v) \in E$ if u and v can communicate to each other. We want to conduct a test on each group such that there is a path between any vertices in that group. In the group testing terminology, for each row j in $M_{t \times n}$, all cells $M[i, j] = 1$ for $i = 1 \ldots n$ must be connected. We call this variant as the connected group testing (CGT) problem. There are several applications for CGT. For example, the multiple-links fault diagnosis problem has been an important and challenging problem in all-optical networks. The most general monitoring structure is to use probing schemes [4]. In particular, probing signals are launched to test the health of the network and the results are used to identify potential failures. The probing signal walks along a set of edges in the graph, thus the probing path must be connected. How to minimize the number of required probing paths is critical to the expense of this technique. Non-adaptive connected group testing can be used to detect d failure edges in the graph with the minimum number of probing paths (which is equivalent to the number of rows in a testing matrix M), thus leading to the new problem: Construct a d-disjunct matrix M so that the graph induced on each row is connected.

References

1. Du D-Z, Hwang FK (2006) Pooling designs: group testing in molecular biology. World Scientific, Singapore
2. Ricciulli L, Lincoln P, Kakkar P (1999) TCP SYN flooding defense. In: Proceedings of CNDS
3. Thai MT, Tiwari R, Du D-Z (2008) On construction of virtual backbone in wireless ad hoc networks with unidirectional links. IEEE Trans Mob Comput 7(8):1–12
4. Harvey NJA, Patrascu M, Wen Y, Yekhanin S, Chan VWS (2007) Non-adaptive fault diagnosis for all-optical networks via combinatorial group testing on graphs. In: Proceedings of IEEE INFOCOM, pp 697–705

Index

M. T. Thai, *Group Testing Theory in Network Security*, SpringerBriefs in Optimization,
DOI: 10.1007/978-1-4614-0128-5,

J

L

M

N

P

R

S

T

V